Why
The Professor
Can't Teach

BOOKS BY MORRIS KLINE

Introduction to Mathematics (co-author), Houghton Mifflin Co., 1937

The Theory of Electromagnetic Waves (editor), Interscience Publishers, 1951

Mathematics in Western Culture, Oxford University Press, 1953

Mathematics and the Physical World, T. Y. Crowell Co., 1959

Mathematics, A Cultural Approach, Addison-Wesley Publishing Co., 1962

Electromagnetic Theory and Geometrical Optics, John Wiley and Sons, 1965

Calculus, An Intuitive and Physical Approach, John Wiley and Sons, 1967

Mathematics for Liberal Arts, Addison-Wesley Publishing Co., 1967

Mathematics in the Modern World (editor), W. H. Freeman and Co., 1968

Mathematical Thought From Ancient to Modern Times, Oxford University Press, 1972

Why Johnny Can't Add: The Failure of the New Mathematics, St. Martin's Press, 1973

Why The Professor Can't Teach

Mathematics and the Dilemma of University Education

by
Morris Kline

Professor Emeritus of Mathematics
Courant Institute of Mathematical Sciences
New York University

St. Martin's Press   New York

Library of Congress Cataloging in Publication Data

Kline, Morris, 1908-
    Why the professor can't teach.

    Bibliography: p.
    1. Mathematics—Study and teaching (Higher)—United
States. I. Title.
QA13.K63   520'.7'1173   76-62777
ISBN 0-312-87867-2

This book is dedicated to the undergraduate students in American colleges and universities.

Contents

Why
The Professor
Can't Teach

Whether 'tis nobler in the mind to suffer
The slings and arrows of outrageous fortune;
Or to take arms against a sea of troubles;
And by opposing end them.

William Shakespeare

Melancholia
by Albrecht Dürer

The Bettmann Archive, Inc.

Preface

There are, indeed, professors ... for whom I have the greatest love and esteem, and think them very capable of instructing youth were they not tied down by established customs. ... Perhaps an attempt may be made some time or other to remove the evil, when it is seen to be not without remedy.

Jean-Jacques Rousseau

This book indicts liberal arts undergraduate education. In view of rather oft-cited failings and shortcomings of education at all levels, why focus on undergraduate education? The liberal arts colleges are the heart of our entire educational system. They address the youth who become our acknowledged leaders or will maintain our cultural level; they offer the prerequisite backgrounds to those who enter the professions of law, medicine, dentistry, and accounting; they provide the basic academic courses for our scientists and engineers; and they educate the primary and secondary school teachers and thereby influence education at these levels. The liberal arts colleges do not explicitly train college teachers, but these men and women tend to perform in the manner in which they have been instructed as undergraduates.

The emphasis in this book is on undergraduate mathemat-

1

ics education. However, much of what will be discussed about the teaching of this subject applies to many other subjects. Numerous conversations with colleagues and a few books that have appeared recently assure me that this is the case.

Even if this were not so, mathematics education is by no means just a matter of teaching one of the three *R*'s. Mathematics is a major branch of our culture, the backbone of our scientific civilization, and the basis of our technology and financial and insurance structures. Its value to the social sciences, biology, and medicine is also by no means negligible. For all students, whether or not they will ever use the knowledge, sound mathematical pedagogy is vital: Failure in mathematics causes them to lose confidence in themselves, and this loss affects their attitude toward all education.

Mathematics education has been a debacle. One need only ask a typical college graduate how much two-thirds of three-quarters is; and if perchance he should give the correct answer, one need only challenge it to discover his uncertainty. The only generally understood fact about the subject is that it is ununderstandable. Adults' reaction, based on schooling in the subject, is a mixture of awe and contempt; usually it is more of the latter. Even otherwise well-educated people boast about their ignorance of the subject as if it were beneath notice.

Why is undergraduate education poor? The prime culprit is the overemphasis on research. The universities justify this emphasis on the grounds that research not only advances knowledge but that proficiency in research is essential to good teaching. I shall examine these contentions and propose to establish that, in today's world, research and undergraduate teaching are in direct conflict. "Publish or perish" is not a threat to professors only, for its actual interpretation is "Publish, and perish the students." To secure research professors, the universities are obliged to offer high salaries, low teaching loads, and the freedom to pursue specialties. Because funds are limited, the universities adopt devious practices to handle other functions, notably undergraduate education.

Although the policies of universities are the root of our educational shortcomings, and professors are subject to these policies, professors cannot be completely exonerated. They have power, a measure of independence if tenured, as most are, and moral obligations to the students. They need not give inordinate attention to research. But professors are all too human. They respond to enticements such as better salaries for research; they compete fiercely for the status accorded to research; and they use that status for material gains—such as royalties from miserably written texts. Nor does their neglect of pedagogy seem to trouble many of them. Professors are content to offer courses that reflect their own values at the expense of student needs and interests. Insofar as these courses are for high school and elementary school teachers, they affect adversely the quality of education at those levels.

There is no panacea for all the ills. But an analysis of what is taking place in our centers of learning points the way to the considerable improvement that could be made with presently available manpower, facilities, and funds.

While the objective arguments offered herein have obvious force and are borne out by statistical data to be found in the bibliographic references, many assertions and recommendations are to some extent personal judgments. But the latter are based on experiences of almost five decades in teaching, research, and administration. For most of that period I have taught on both the undergraduate and graduate levels, and I have served for eleven years as chairman of undergraduate mathematics at a major center. As for research, I spent two years as a research assistant at the Institute for Advanced Study, over three years in war research, and twenty years as director of a research division in one of the best graduate mathematics departments in this country. Numerous invited lectures at other colleges and universities and a few visiting professorships have also enabled me to become directly informed about what these institutions are doing.

Criticism of university policies and practices is by no means new. Many of the books on higher education in America discuss poor management, lack of educational effectiveness, waste, and the gulf between administration and faculty. Unfortunately, the authors seem to believe that objectivity necessitates wishy-washiness; they argue pro and con and present vague opinions about what might be done. One gets the impression that the authors are afraid to draw conclusions or to come to grips with the real issues. Perhaps the reason for the fence-straddling is the timidity ascribed to academics—or perhaps it is an honest failure to recognize priorities. However, we cannot in good conscience continue to vacillate, tread softly, and settle for bland statements. Millions of students pass through the universities, and their lives are seriously affected by that experience. It is true that the universities have many obligations and roles in our society, but the education of liberal arts students is undeniably the fundamental one, and any equivocation about the primacy of this function is an abdication of responsibility.

I hope that this book will serve several groups of people.

University administrators who are concerned with the effectiveness of their organizations should reconsider university priorities and policies.

Parents, many of whom sacrifice even some necessities of

life to send their children to college, ought to be interested in what treatment their children may receive; the prestige of a university is no assurance of a sound educational process.

Legislators who apportion funds for education should be informed on how they are used.

Even research professors cannot stand aloof. The greatest threat to the life of mathematics is posed by the mathematicians themselves, and their most potent weapon is their poor pedagogy. Mathematics as a part of liberal arts education may disappear just as Greek and Latin have disappeared and as modern foreign language studies are now gradually disappearing.

Finally, citizens who believe that the education of our people is the surest guarantee of the effectiveness of a democratic government should concern themselves with the operation of our key educational institutions.

I wish to thank Professor Wilhelm Magnus of the Polytechnic Institute of New York, Professor George Booth of Brooklyn College, City University of New York, and Professor Murray S. Klamkin of The University of Alberta for their critical reading of the manuscript. Their willingness to help in no way implies their endorsement of the contents of the book. I am also deeply indebted to Miss Julie Garriott for her thorough and thoughtful editing and for other contributions to the publication, which extended far beyond her obligations. Lastly grateful acknowledgment to my wife Helen for her critical reading of the manuscript and her punctilious proofreading.

Morris Kline
June 1977

1

The Vicious Circle

In a self-centered circle, he goes round and round,
That he is a wonder is true;
For who but an egotist ever could be
Circumference and center, too.

Sarah Fells

Peter Landers found himself caught in a vicious circle. He had just secured a Ph.D. in mathematics from Prestidigious University and, having been well recommended, readily secured a faculty position at Admirable University. Thereupon Peter faced the problem of teaching mathematics to prospective engineers, social scientists, physicists, elementary and secondary school teachers, the general liberal arts students, and those who, like himself, had chosen to become mathematicians. Peter was fully aware of these varied career interests, and he also knew that students came to college with different drives and preparation. But he was confident that his education, typical for Ph.D.'s, had prepared him for the tasks ahead.

To put himself in the proper frame of mind he reviewed his own education. The elementary school courses had been acceptable. After all, one did have to know how much to pay for five candy bars if he knew the price of a single bar. True, some operations were baffling. It had not been clear why the division of two fractions had to be performed by inverting the denominator and multiplying—but the teacher seemed to know what was correct. He had constantly referred to rules, principles, and laws. Rules, like rules of behavior, apparently applied to arithmetic, too. For all Peter had known, principles were laid down by the principals of the schools, and certainly *they* were authorities. As for laws, everyone knew that there were city laws, state laws, federal laws, and even the laws of the Ten Commandments. Certainly laws must be obeyed. Though under some tension as to whether he was violating laws, Peter was young and resilient. In any case, what to do was clear and the answers were right.

In his review of his high school education Peter did recall some doubts he had had about the value of what he was being taught. He hadn't understood why the teacher had to stress that the sum of two whole numbers is a whole number, or why he had to prove that there is one and only one midpoint on every line segment; but evidently the teacher was trying to make sure that no one could be mistaken on these elementary matters. After all, teachers knew best what had to be done.

Peter also recalled one teacher's enthusiasm about the quadratic formula. "You see," the teacher proclaimed triumphantly after he had derived the formula, "we can now solve any quadratic equation." But Peter had been perverse and had asked the teacher why anyone wanted to solve any quadratic equation. The teacher's reply was a disdainful look that caused Peter to shrink back. His question must have been a silly one.

He remembered a similar experience in geometry. After a long and apparently strenuous effort, the teacher proved that two triangles are congruent if the sides of one are equal respectively to the sides of the other. Then he turned to the class as if expecting applause. Again Peter dared to speak up: "But isn't that obvious? A triangle is a rigid figure. If you put three sticks together to form a triangle, you cannot change its size or shape." Peter had learned this at the age of five while playing with Erector sets. The teacher's contempt was obvious. "Who's talking about sticks? We are concerned with triangles."

Despite a few other disagreeable incidents Peter continued to like mathematics. He believed in his teachers. It was easy to comply with their requests, and the certitude of the results gave him, as they had given others before him, immense satisfaction. And so Peter moved on to college with the conviction that he liked mathematics and was going to major in it.

His first experiences were disturbing. After his program was approved by an adviser who did not understand what an Advanced Placement Examination Grade of 4.5 meant—the adviser had thought that 10 was a perfect grade so that 4.5 was a poor one—Peter was finally registered.

He entered his first college classroom for a course which happened to be English. To his surprise he found about five hundred students already seated. The professor arrived, delivered his lecture, and, obviously very busy, rushed out of the room. Peter never found out what his name was, but apparently names were not important, because the professor never bothered to ask any student his name either. Nor, Peter thought, would the professor have noticed had a different group of five hundred students appeared each time. Term papers were required, and these were graded by graduate students who insisted that "Who shall I call next?" was correct, though Peter had been taught otherwise in high school. The size of the class and the impersonal character of the instruction disturbed Peter at first, but he soon realized that the requirements of the English course could be met merely by listening. And so he relaxed.

Peter's second class, one in social science, surprised him for different reasons. At the professor's desk was a young man not much older than Peter. As the instructor conducted the lesson he was obviously nervous. Somehow the lessons throughout the semester were confined almost entirely to the first part of the text. And the instructor did not welcome questions.

The third class—mathematics—was a shock. Peter entered the room and found that it was a large auditorium. At the bottom of the room, at the professor's desk, was not a man but a box, which proved to be a television set. Shortly after Peter's entrance the box began to speak and the students took notes feverishly. From many seats one could not see clearly, if at all. But by coming early one could get a good seat. And so Peter managed to learn some of his college mathematics by listening and looking at a TV program.

Though it was not a requirement, Peter decided to take some physics. He had heard somewhere that mathematics was applied to physics, and he thought he should find out what these applications were. The physics professor constantly talked about infinitesimals and which infinitesimals could be neglected. The mathematics professors, however, had warned that such concepts and procedures were loose and even incorrect. But Peter listened attentively. He was sure that even though the mathematics and the physics professors apparently did not communicate with each other and so did not talk the same language, their methodologies could be reconciled. He did seek counsel from his professors on this matter, but unfortunately they were not available. One was actually living out of the city, in Washington, D.C.; another was always involved in consultations outside the university; and a third had office hours only on Sundays, from 6:00 to 8:00 A.M.

In the junior and senior years the classes were smaller, and the courses were usually taught by older faculty. Many blithely ignored the texts they had assigned and spent the period transferring material from their notes to the board. The professors copied assiduously and the students did likewise. When the professors looked up from their notes they looked into the blackboard as though the students were behind it.

Nevertheless Peter persevered, received his bachelor's degree, and proceeded to graduate school. His experiences there paralleled those of most other students. Professors were hard to contact. The bulletin descriptions of the courses bore no relation to what the professors taught. Each professor presented his own specialty as though nothing had been done or was being done by anyone else in the world. And so Peter learned about categories, infinite Abelian groups, diffeomorphisms, noncommutative rings, and a variety of other specialties.

Prospective Ph.D.'s must write a doctoral thesis. Finding a thesis adviser was like hunting for water in a desert. After many trials, including writing theses on topics suggested by his professor that, it turned out, had been done elsewhere and even published, Peter wrote a thesis on almost perfect numbers that completed his work for the degree.

With the Ph.D. behind him, Peter presumed he was prepared for college teaching. Upon taking up his position at Admirable University he received from his department chairman the syllabi for the several courses he was to teach and was told what texts he was to use for these courses. Cheerful, personable Peter went about his assigned tasks with enthusiasm. He had always liked mathematics and had no doubt that he could convey his enthusiasm and understanding of the subject to his students. He had been informed by the chairman that to secure promotion and tenure he would be expected to do research. This requirement in no way dimmed Peter's spirit, because he had been told repeatedly that mathematicians do research and was confident that the training he had received had prepared him for it.

But the world soon began to close in on Peter. As a novice he was assigned to teach freshmen and sophomores. His first course was for liberal arts students, that is, students who do not intend to use mathematics professionally but who take it either to meet a requirement for a degree or just to learn more about the subject. Recognizing that many of these students are weak in algebra, Peter thought he would review negative numbers. To make these numbers meaningful he reminded the students that they are used to represent temperatures below zero; and to emphasize the physical significance of negative temperatures he pointed out that water freezes at 32° F., so that a negative temperature means a state far below freezing. Though the example was pedagogically wise, Peter could see at once that the students' minds had also frozen, and the rest of his lesson could not penetrate the ice.

In a later lesson Peter tried another subject. As an algebraist by preference he thought students would enjoy learning about a novel algebra. There is an arithmetic that reduces all whole numbers by the nearest multiple of twelve. To make his lesson concrete Peter presented clock arithmetic as a practical example: Clocks ignore multiples of twelve, so that four hours after ten o'clock is two o'clock. The mere mention of clocks caused the students to look at their watches, and it was obvious that they were counting the minutes until the end of the period.

And so Peter tried another novelty, the Koenigsberg

bridge problem. Some two hundred years ago the citizens of the village of Koenigsberg in East Prussia became intrigued with the problem of crossing seven nearby bridges in succession without recrossing any. The problem attracted Leonhard Euler, the eighteenth century's greatest mathematician, and he soon showed by an ingenious trick that such a path was impossible. The villagers, who did not know this, continued for years to amuse themselves by making one trip after another during their walks on sunny afternoons—but when Peter presented the problem in the artificial, gloomy light of the classroom, a chill descended on the class.

Peter's next class was a group of pre-engineering students. These students, he was sure, would appreciate mathematics, and so he introduced the subject of Boolean algebra. This algebra, created by the mathematician and logician George Boole, does have application to the design of electric circuits. The mention of electric circuits appeared to arouse some interest, and so Peter explained Boolean algebra. But then one student asked Peter how one uses the algebra to design circuits. Unfortunately, Peter's training had been in pure mathematics and he did not know how to answer the question. He was compelled to admit this and detected obvious signs of disappointment and hostility in the students. They evidently believed that they had been tricked. In his attempts to explain and clarify other mathematical themes Peter also learned that engineering students cared only about rules they could use for building things. Mathematics proper was of no interest.

Nor were the premedical students any more kindly disposed to mathematics. Their attitude was that doctors do not use mathematics but take it only because it is required for the physics course, and even the physics seemed of dubious value. The physical and social scientists had a similar attitude. Mathematics was a tool. They were interested in the real world and in real people, and certainly mathematics was not part of that reality.

Peter was soon called upon to teach prospective elementary and high school teachers. He did not expect much of the former. These students were preparing to teach many different subjects and so could not take a strong interest in mathematics. However, high school teachers specialize in one area, and Peter certainly expected them to appreciate what he had to offer. But every time he introduced a new topic, the first question the students asked was, "Will we have to teach this?" Peter did not know what the high schools were currently teaching or what they were likely to teach in any changes impending in the high school curriculum. Hence, he honestly answered either "No" or "I don't know." Upon hearing either response the prospective teachers withdrew into their shells, and Peter's teachings were reflected from impenetrable surfaces.

Peter's one hope for a response to his enthusiasm for teaching was the mathematics majors. Surely they would appreciate what he had to offer. But even these students seemed to want to "get it over with." If he presented a theorem and proof, they noted them carefully and could repeat them on tests; however, any discussion of why the theorem was useful or why one method of proof was likely to be more successful or more desirable than another bored them.

A couple of years of desperate but fruitless efforts caused Peter to sit back and think. He had projected himself and his own values and he had failed. He was not reaching his students. The liberal arts students saw no value in mathematics. The mathematics majors pursued mathematics because, like Peter, they were pleased to get correct answers to problems. But there was no genuine interest in the subject. Those students who would use mathematics in some profession or career insisted on being shown immediately how the material could be useful to them. A mere assurance that they would need it did not suffice. And so Peter began to wonder whether the subject matter prescribed in the syllabi was really suitable. Perhaps, unintentionally, he was wasting his students' time.

Peter decided to investigate the value of the material he had been asked to teach. His first recourse was to check with his colleagues, who had taught from five to twenty-five or more years. But they knew no more than Peter about what physical scientists, social scientists, engineers, and high school and elementary school teachers really ought to learn. Like himself, they merely followed syllabi—and no one knew who had written the syllabi.

Peter's next recourse was to examine the textbooks in the field. Surely professors in other institutions had overcome the problems he faced. His first glance through publishers' catalogues cheered him. He saw titles such as *Mathematics for Liberal Arts, Mathematics for Biologists, Calculus for Social Scientists,* and *Applied Mathematics for Engineers.* He eagerly secured copies. But the texts proved to be a crushing disappointment. Only the authors' and publishers' names seemed to differentiate them. The contents were about the same, whether the authors in their prefaces or the publishers in their advertising literature professed to address liberal arts students, prospective engineers, students of business, or prospective teachers. Motivation and use of the mathematics were entirely ignored. It was evident that these authors had no idea of what anyone did with mathematics.

Clearly a variety of new courses had to be fashioned and texts written that would present material appropriate for the respective audiences. The task was, of course, enormous, and it was certain that it could not be accomplished by one man over a few years' time. Nevertheless Peter became enthusiastic about the prospect of interesting investigations and writing that would lure students into the study of mathematics and endear it to them. The spirit of the teacher arose and swelled within him. As these pleasant thoughts swirled through his mind, another, dampening thought, like a dark cloud on the horizon, soon entered. He was a recently appointed professor. Promotion and, more important, tenure were yet to be secured. Without these his efforts to improve teaching would be pointless—he would be unable to put the product of his work to use. But promotion and tenure were obtained through research in some highly advanced and recondite problems almost necessarily chosen in the only field in which he had acquired some competence through his doctoral work. Such research was no minor undertaking. It demanded full time and total effort.

Clearly, he must give the research precedence, and then perhaps he could undertake the improvement of teaching. And so for practical reasons Peter decided to devote the next few years to research. But the struggle to publish and to remain in the swim for promotion and salary increases caught Peter in a vortex of never-ending spirals of motion; and the closer he came to the center the deeper he was sucked into research. In the meantime Peter continued to teach in accordance with the syllabi and texts handed down to him by his chairman. His few, necessarily limited efforts to stir up some activity among his older colleagues, who were in a better position to break from the existing patterns, were futile because these professors had accepted the existing state of affairs and chose to shine in research. Success there was more prestigious and more lucrative.

Ultimately, Peter, like other human beings, succumbed to the lures that prominence in research held forth. As for the students—well, students came and went, and they soon became vague faces and unremembered names. Education might hope for an epiphany, but Peter was not ordained to be the god of educational reformation. By the time he had acquired tenure he had joined the club. Like others before him he concentrated on research and the training of future researchers who would also be compelled to resort to perfunctory and ineffective teaching. Peter had taken his place in the vicious circle.

The history of Peter Landers' aborted teaching efforts, real enough, seems exaggerated. One might conceive of its taking place in nineteenth-century Germany or France. But the United States is devoted to education. We were the first nation to espouse universal education and to foster the realization of the potential of every youth. Our Founding Fathers, notably Benjamin Franklin and Thomas Jefferson, stressed the necessity of this policy, and it was adopted. Even today no country matches the educational opportunities and facilities that the United States provides for its youth. But the practices within educational institutions seem to be in marked variance with the principles and policies of our country.

How has it come to pass that Peter and the many thousands of his colleagues find themselves enslaved by research, while education, the major goal of our vast educational system, is being sacrificed? Does the pressure to do research stem from the professors because they prefer the prestige and monetary rewards? Or does it come from the university administrations? In either case, does not research make for better teaching? Or is there a conflict between the two, and if there is, how can we resolve it? Since the crux of the problem lies with the universities—which train the teachers of all educational disciplines and at all levels—we must examine the policies and practices of our higher educational institutions.

2

The Rise of American Mathematics

It must be observed, I do not esteem, as public institutions, those ridiculous establishments that go by the name of Universities.

Jean-Jacques Rousseau

Clearly, the relationship between teaching and research—their compatibility or incompatibility, their possible mutual reinforcement or opposition—must be investigated. To carry out these tasks we shall first examine the rise and present status of American education. In particular we must look into the role that research has come to play. This history will at least tell us why we are where we are.

One of this country's marks of greatness, as Ralph Waldo Emerson pointed out in *Education*, is that schooling at many levels was undertaken almost immediately by the first settlers. The greatness is underscored by the fact that the United States was founded in the main by poor, uneducated people of vastly different backgrounds and languages, people who came here to improve their lot, whether by gaining political freedom, spiritual freedom, or sheer material necessities. It is true that many sought only the freedom to practice their own brand of religious intolerance. However, no matter what values were sought, education was undertaken at the very outset.

Elementary schools, set up at once, were soon followed by secondary schools. The earliest of the latter were private Latin grammar schools, the first of which, the Boston Latin Grammar school, was founded in 1635. Even colleges were established early. Harvard College opened its doors in 1636 and William and Mary was next in 1693. Yale started in 1701 as the Collegiate School of Connecticut in Old Saybrook, and then moved to New Haven in 1716 at which time it adopted the name of its benefactor, Elihu Yale.

The numerous colleges founded in the seventeenth and eighteenth centuries were sectarian and in fact at first were devoted to supplying clergymen for the several faiths. Harvard, for example, trained Puritan ministers. The only nondenominational school in the colonies up to 1765 was the University of Pennsylvania. The policy of separation of church and state, which was adopted by the Republic in 1787, made public support of sectarian schools illegal. Even where indirect support might have been possible, religious differences and animosities caused legislators to refuse public support to sectarian colleges of faiths adverse to their own. Hence, many nonsectarian colleges were founded. The Morrill Act, passed by Congress in 1862, marked a major turning point by supplying public funds to "godless" universities for "the Benefit of Agriculture and the Mechanic Arts." This Act enabled states to found the land-grant institutions, among which the Midwestern universities became leaders. Many more public and private colleges and universities were established in the succeeding decades, with church affiliation still the more common. As late as 1868, Cornell opened amid much criticism because it had no sectarian ties.

To be sure, education was not widespread at the outset. In the seventeenth century only one colony, Massachusetts, insisted on a few years of compulsory education for all its children. As the country became somewhat settled this practice spread, and by 1910 all but six states had adopted it. Concurrently the age to which students had to remain in school was raised to fourteen, sixteen, and ultimately eighteen, though it differed from state to state. It is remarkable that in the early part of the nineteenth century the United States was the only country with some compulsory education.

Moreover, by the late nineteenth century the right to free primary and secondary education was recognized, though this right was not widely utilized. In 1890 only 7 percent of the children of high school age actually attended schools, and only about 3 percent of the appropriate age group attended a college or university. As the population grew not only the number of students, but also the percentages increased rapidly. The right of high school graduates to low-cost college education was granted first by the big Midwestern universities. The privilege was specious, however, because these universities admitted all high school graduates and then flunked 50 to 70 percent in the first year. Nevertheless, the right to a college education gained ground all over the United States.

What was the quality of the education? In colonial times and in the early decades of the Republic, because food, shelter, and clothing had to be given precedence, education amounted to little more than the transmission of ignorance. Certainly the pursuit of mathematics and the physical sciences, which might contribute in the long run even to the increase of material goods, had to be sacrificed to immediate needs. The country concentrated on reading, writing, arithmetic, and religion. Pedagogy, which had already been receiving some attention in Europe, received no attention here for a couple of centuries: primers showing the common man how to do simple arithmetic can hardly be dignified with that term, especially since copying numerals and rote counting were the most that was taught.

The need for arithmetic in commerce, exploration, surveying, and navigation motivated the introduction of *full-fledged* courses in that subject—but surprisingly, not in the elementary schools. In fact it was not until the eighteenth century that it was taught and, if at all, in the colleges and universities. Before 1729, the arithmetic texts were reprints of texts from England. After that date such texts were written and published in America. (The first one was written by Isaac Greenwood, a professor at Harvard College from 1728 to 1738, who had the advantage of studying in England and the disadvantage of being fired for intemperance.)

The rest of the eighteenth-century curriculum was a hodgepodge of cultural materials and moral imperatives. Keeping young people within the fold was a major concern; preparation for careers, the ministry, law, and medicine continued to be important. To this was added education in agriculture and forestry. A century later the land-grant colleges were founded to teach the latter two subjects specifically, though they soon expanded to include academic work. Utility and personal success, Alexis de Tocqueville observed in his *Democracy in America* (1835), were the chief concerns. Gradually the colleges and universities adopted liberal arts subjects with an emphasis, following the English model, on the classics. Molding character and teaching an aristocratic style of life to the well-born were also objectives, at least until 1900.

During the eighteenth century arithmetic was gradually moved down into the Latin grammar schools and high schools, where it and the other subjects were taught mainly as preparation for college. In turn the colleges—Harvard, William and Mary, Yale, Princeton, Pennsylvania, and others—began to require arithmetic for admission. Yale did so in 1745 and Princeton in 1761; however, Harvard did not do so until 1803.

In the nineteenth century arithmetic finally became an elementary school subject, mainly because the growing industrialization of the country required more knowledge from workers. But mental discipline was also a reason for teaching the subject. Elementary schools in Massachusetts and New Hampshire were the first to teach arithmetic. This was in 1789. Only by the end of the nineteenth century was arithmetic firmly a part of the elementary school curriculum.

The increasing need for mathematics in manufacturing, railroading, engineering, cartography, and the study of science—especially mechanics and astronomy—motivated the introduction of algebra, geometry, and trigonometry. These subjects entered the curriculum on the college level. They were taught as junior and senior courses at Harvard, Yale, and Dartmouth from 1788 on and were retained by most colleges until the end of the nineteenth century. Thus, the mathematics curriculum at the University of Pennsylvania in 1829 reviewed arithmetic and taught the beginnings of algebra and geometry in the freshman year. The sophomore year covered more of algebra and geometry, some plane and spherical trigonometry, surveying and mensuration. Only in the third year were what are now beginning college subjects, such as analytic geometry and differential calculus, introduced, along with perspective and mathematical geography. The senior year completed elementary calculus and initiated work in the essentially mathematical subjects of dynamics and astronomy.

Early in the nineteenth century a few colleges began to

require for admission some of what we now call high school mathematics. Thus in 1820 Harvard upgraded its admission requirements to include elementary algebra. Yale did so in 1847 and Princeton in 1848. As for Euclidean geometry, Yale in 1865 was the first to require it for admission. Princeton, Michigan, and Cornell followed suit in 1868 and Harvard in 1870. During the last part of the nineteenth century algebra and geometry finally became high school subjects. Preparation for college, as colleges gradually began to require algebra and geometry for admission, and mental discipline were stressed as the reasons for teaching these subjects on the high school level.

From 1820 to 1900 the universities gradually raised their level of mathematics instruction. By 1900 trigonometry, analytic geometry, and the calculus became standard college subjects. Although most colleges went no further in the early part of the nineteenth century, a few moved on to differential equations and more advanced subjects. Nevertheless,the general level of mathematical knowledge was low.

The status of American mathematics in 1900 can be judged by an incident that is ludicrously revealing. A country physician, Edward J. Goodwin, who possessed no sound knowledge of mathematics, submitted a bill to the Indiana Legislature in 1897 that called for declaring the value of π to be 4. A comparison of the common formulas for the area of a circle and the area of a circumscribed square shows immediately that π cannot be 4. Nevertheless, the Legislature considered the bill. Fortunately the Senate, thoroughly modern in one respect, postponed action and the bill was never passed. One may not be too surprised by the actions of elected officials, but the *American Mathematical Monthly,* a journal founded by leaders of that time, in its first volume,

July 1894, published Goodwin's proposal and in 1895 dutifully printed other absurd Goodwinisms.

The pressure in the United States to raise the levels of mathematics education and to educate more and more of American youth increased sharply as the country became a greater world power. World War I certainly showed the need for more mathematics education, and ever-increasing technological uses of mathematics added to the pressure. The universities responded by raising the requirements for admission and by adding advanced courses.

This brief sketch of the rise of mathematics education has not addressed the question of how teachers were procured. The upgrading of the formal level of education and the increase in the number of students did not in itself provide a supply of teachers. In colonial times almost anyone could set himself up as a teacher, and since the population of this country was made up mainly of uneducated people, one can readily appreciate the level of teaching. As late as 1830, Warren Colburn, one of the early-nineteenth-century American educators, said in an address:

> The business of teaching, except in great seminaries, has not been considered as one of the most honorable occupations, but rather degraded; so that few persons of talent would engage in it. Even in our own country and age, it has been too much the case that persons with a little learning, and unwilling to work and unfit for anything else, have turned schoolmasters, and have been encouraged in it. They have been encouraged in it because the pay of school teachers, in most instances, has been just sufficient to obtain that class of persons . . . .

Although the colonial colleges were not founded to train teachers, many did so. But the colleges admitted only men whereas, because of the low salaries, women were the most likely candidates.* Also, most colleges were still sectarian and so could not be the main source of public school teachers or be favored by public money for the education of teachers. Further, the better colleges insisted on more qualifications for admission than prospective school teachers could afford to obtain.

Recognition of these facts prompted Horace Mann, in the 1830s, to advocate special means for training teachers. State normal schools were founded that required only an elementary school education, gave teachers a modicum of training, and then sent them out to teach. These schools were organized under Mann's leadership in the late 1830s and they spread throughout the country.

The problem of supplying teachers for a public school system that kept extending universal compulsory education was never really solved. It was aggravated by the fact that more and more immigrants, uneducated and poor, entered the country. Nor was the situation any better in the colleges. When Charles William Eliot became president of Harvard College in 1869, he emphasized that one of the outstanding problems was to induce ambitious young men to adopt the calling of professor. "Very few Americans of eminent ability," he said, "are attracted to this profession." The supply of good teachers continued to fall far short of the need. Certainly many who entered teaching in the nineteenth and early twentieth centuries were not the best choices. Teaching remained very poorly paid, had little or no prestige, and required few qualifications. Of course, some capable people, unable to afford the education that other professions required, turned to teaching.

The colleges and universities, which were not themselves

---

*   Oberlin College did begin the admission of women in 1837, but few others did so until after the Civil War.

well staffed, gradually took over the function of educating teachers. Professors of education became members of college faculties beginning in 1832 at New York University and at other colleges soon after.

Up to 1920, however, the training of elementary school teachers was still done in the normal schools, which by that time required a two- to four-year course of study. The colleges trained the nation's high school teachers. Then the colleges and universities began to establish departments or schools of education to train both elementary and high school teachers. Beginning around 1950, the normal schools were converted to four-year liberal arts colleges with emphasis on teacher training. High school mathematics teachers now take a four-year liberal arts course with a major in mathematics, but elementary school teachers take little academic mathematics and a great deal of instruction in pedagogy.

The rise of institutions and faculties devoted to teacher education did not improve the knowledge of mathematics proper that the professors should have possessed. In fact, the level of mathematics instruction was very low until well into the twentieth century. American texts were poor; good ones had to be imported.

A few people educated themselves. One of the outstanding examples was Nathaniel Bowditch (1773-1838), who translated and added explanatory notes to Laplace's *Mécanique céleste,* one of the great works of European mathematics. This translation was the first substantial mathematics book in the United States and, in fact, in the entire Western hemisphere. The first truly great American scientist was the self-taught physical chemist Josiah Willard Gibbs (1839-1903), who is also known for his contribution to vector analysis. Many professors went to Europe for their education; these men were certainly the best educated and were actually the only ones qualified to teach mathematics and train mathematicians.

Gradually the quality of education and available books improved sufficiently for the United States to train its own mathematicians, though hardly in numbers sufficient to serve the needs of the country. The first of the outstanding American-trained mathematicians was Benjamin Peirce (1809-1880), a Harvard graduate and professor at Harvard from 1831 to his death. Other Harvard mathematics professors of his time, such as William Fogg Osgood and Maxime Bocher, got their Ph.D.'s in Europe before joining the university faculty.

Despite the increase during the early decades of this century in well-trained mathematicians, one could not be complacent about mathematics education. There were too few teachers to staff the increasing number of elementary schools, high schools, colleges, and universities. The states, cities, and towns lagged in their appreciation of the value of education and were miserly in funding. Consequently, salaries remained low and potentially fine teachers continued to choose other jobs and professions.

Some indication of the quality of the teachers comes from data compiled by the Educational Testing Service. As late as 1954 the elementary school teachers who were interviewed feared and hated mathematics. Naturally this influenced their teaching. Half of 370 teachers tested could not tell when one fraction was larger than another. Thus, the teachers knew less than they were required to teach. High school teachers in most states could qualify for a license to teach mathematics with only ten hours of college mathematics. In many states a license to teach in high school is still sufficient to qualify for teaching any subject.

Since until recently ignorance was the outstanding characteristic of American educators at all levels, not only was the factual content of mathematics courses low, but also the educational goals were either imperfectly or mistakenly perceived. We need not pursue here the conflicts, disagreements, and theories advanced over the several centuries of American education. That social utility should be an objective, particularly when the colonies and even the young Republic were struggling for survival, could hardly be challenged; but arithmetic was also advocated as a mental discipline that would extend to other areas. Though portions of secondary school mathematics could be defended on the ground of their usefulness in applications such as surveying and navigation, it was impossible to defend most of high school mathematics on this ground. Instead, mental discipline, knowledge of a branch of our culture, the inherent beauty of the subject, and preparation for college were advanced as justifications.

As to the pedagogy, intuitive approaches using concrete materials and sensory experiences, espoused, for example, by Johann Heinrich Pestalozzi (1746-1827), and abstract rigorous approaches were both advocated to impart the values and even the meaning of mathematics. In practice, however, the teachers, barely understanding what they were taught, handed down the processes and the proofs mechanically. Drill was the order of the day. Any questions were answered with the dogmatic reply: This is the way to do it. To divide one fraction by another, invert and multiply. The product of two negative numbers is a positive number. If $x + 2 = 7$, transpose the 2 and change the sign so that $x = 7 - 2$, or 5. The rationale of geometric proofs was never given. Students memorized them and handed them back on examinations.

The departments and colleges of education attached to universities were not helpful. The reasons are patent. The mathematics educators were themselves taught subject matter in the same way as others were taught, and so their understanding of mathematics was no better than that of good students. It was in fact worse, because the educators did not deem it necessary to go as far in their subject as those who planned to be professional mathematicians and scientists. Furthermore, very little was known then (and even now) about the psychology of teaching and learning. Hence, the educators had little to contribute. They taught prospective teachers how to carry out the drill and memorization, the very processes by which they had been taught.

Another unproductive effort was to call upon psychologists, who commonly offered a standard course in the psychology of education. But the psychologists also had little to offer. Professor Edward Lee Thorndike (1874-1949) of Columbia University's Teachers College advocated massive repetition or drill. Students should be trained to respond automatically: 3 + 2 should immediately elicit 5. Understanding would come eventually. But this advice was no more than an endorsement of the rote teaching that had been practiced for generations. (See also Chapter 9.) This method of teaching is still advocated by educators such as Professor B.F. Skinner of Harvard—now it is called programmed learning.

To improve education, innumerable commissions and committees were appointed by mathematical organizations and state and local governments. The committees were to study goals, content, and pedagogy. The history is extensive but irrelevant. One factor that hampered the reforms suggested by the many committees, conferences, symposia, and commissions is that no sound evaluation was employed to test the recommendations. Another was that changes were introduced too quickly, making it difficult for teachers to implement them properly. The efforts and sincerity of the mathematics educators are not in question. However, the nature and goals of mathematics education, including appropriate subject matter, the best methods of teaching it, the role of applications, and the means to attract and involve students were not determined by these studies.

Undoubtedly, changing conditions in this country rendered pointless many recommendations that may have been right in their time. The proportion of students attending high school in the United States grew from 12 percent in 1900 to over 90 percent in 1967. Obviously the interests and backgrounds of high school students became far more varied as that population increased. The Bachelor of Arts (or Science) curriculum, designed originally to give American youth some knowledge of classical (Greek) and European culture, was broadened somewhat after the Civil War to include the social, physical, and biological sciences and the humanities. In 1900, when about 4 percent of the college-age group went to college, this program may have been proper. In 1970 about 48 percent, or seven million students, went to college, for totally different reasons than just the acquisition of knowledge. The colleges had to adapt to many more levels of academic ability, far more diverse backgrounds, and many new goals—notably career training in a wide variety of fields. Beyond these changes was the initiation in 1901 of community or two-year colleges, which were the successors of the private junior colleges first established in the late nineteenth century and which were preparatory for the senior colleges. The community colleges now have many terminal students who seek essentially a vocation.

Despite gradual improvement in the quality of the teaching staff on the lower educational levels, a stabilized educational system, reasonable facilities, and the far greater availability of texts and books, mathematics education was on the defensive from 1920 to 1945. Students did poorly. Enrollment in academic mathematics courses decreased, notably in algebra and geometry. The transfer of training was, perhaps rightly, deprecated. During this period the colleges reduced requirements in mathematics for admission, and students took fewer mathematics courses once enrolled. Many institutions even dropped the mathematics requirement, and many high schools also dropped mathematics as a requirement for a diploma. The value of any mathematics education was widely challenged.

The status of mathematics in the 1930s was described by mathematics professor Eric T. Bell in the *American Mathematical Monthly* of 1935:

> Are not mathematicians and teachers of mathematics in liberal America today facing the bitterest struggle for their continued existence in the history of our Republic? American mathematics is exactly where, by common social justice, it should be—in harnessed retreat, fighting a desperate rear-guard action to ward off annihilation. Until something more substantial than what has yet been exhibited, both practical and spiritual, is shown the non-mathematical public as a justification of its continued support of mathematics and mathematicians, both the subject and its cultivators will have only themselves to thank if our immediate successors exterminate both.

The subsequent history and status of mathematics education was affected by the entry of a new factor—research, in the sense of original contributions. In the area of mathematics this necessarily came rather late. Knowledge in this subject is cumulative. To contribute to mathematics in 1800, for example, one had to know the work of the Greeks, Descartes, Fermat, Newton, Leibniz, Euler, Lagrange, Laplace, and many others. In science the situation was roughly similar. Mechanics was well developed by the end of the eighteenth century, and contributions by neophytes could hardly be expected. The newer fields of science offered more opportunities. In electricity, for example, which was just beginning to be cultivated in 1800, it was possible for the American Joseph Henry to share honors with the Britisher Michael Faraday in the discovery of electromagnetic induction. In the nineteenth century, however, American science as a whole was, like American mathematics, below the European level.

Researchers in general had no place in the United States before about 1850. They could become teachers at low-level colleges and universities and spend their spare time, of which there was little, in research. But such endeavors received no encouragement. In fact, a research person might even arouse suspicion that he was not paying enough attention to his teaching and might even have radical ideas about curriculum and the method of instruction. In 1857, a committee of the Columbia College Board of Trustees attributed the poor quality of the college's educational efforts to the fact that the professors "wrote books." A professor's obligation was not to advance knowledge but to transmit it.

Of course the United States at that time was not really prepared to do research, much less train Ph.D.'s for research. Though during the nineteenth century, as we have already noted, many American professors went to Europe to study, their number was relatively small. Moreover, while a few years in a great cultural center helps immensely, it does not produce researchers, particularly if they return to an intellectual atmosphere in which research is neither cultivated nor widely appreciated. Researchers need the stimulus of colleagues with common interests and of students who carry on the work of the masters and press them for problems and aid.

In effect, during the nineteenth century good understanding and appreciation of research did not exist. For example, Josiah Willard Gibbs, the physical chemist mentioned earlier, worked at Yale University, where for many years he received no pay. In Europe his research in thermodynamics was well understood and highly valued. He wrote a first-class book on statistical mechanics in 1901 that was appreciated by such masters as Felix Klein and Henri Poincaré but ignored in the United States. When the distinguished German mathematician, physicist, and physician Hermann von Helmholtz visited Yale in 1893 and was greeted by the top dignitaries of the university, he asked where Gibbs was. The university officials, nonplussed, looked at each other and said, "Who?"

The state of research in the United States even in the early twentieth century may be illustrated by an incident in the life of Walter Burton Ford, an American mathematician who died at the age of ninety-seven in 1971. He submitted his doctoral thesis to Harvard. The paper was judged by Bocher, Osgood, and Byerly, who were leaders in American mathematics in the first few decades of this century, and deemed unacceptable. Ford sent the paper to the reputable French *Journal de mathématiques,* where it received praise from the editors and was published. The Harvard faculty then reversed its decision and awarded the Ph.D. to Ford.

Another of the early great American mathematicians was Norbert Wiener (1894-1964). His fields of endeavor were not understood here; however, when the European mathematicians began to praise Wiener's work, the American mathematicians took notice and fortunately, if belatedly, he received honor here.

When research in mathematics and the sciences was first undertaken, about 1850, the university leaders thought that it belonged not in the universities but in special institutions referred to as academies. Advanced degrees were offered only in law, medicine, and theology. Yale, however, did initiate graduate training in 1847 and conferred the first Ph.D. degree in mathematics in 1861. Harvard instituted a graduate program in 1872 and conferred the first degree in mathematics in 1873.

Though the universities did begin to undertake research and training for research, the inadequacy of the American research capacity was recognized by some men who had become aware of the high standards of the German universities. It was these men who persuaded wealthy individuals to found universities that would stress research and strengthen the already existing advanced training. David Coit Gilman induced Johns Hopkins, a merchant and banker, to found the institution named after him, which opened in 1876. Leland Stanford, who made his fortune in railroading, established Stanford University, which opened in 1891. And William Rainey Harper convinced John D. Rockefeller to finance the University of Chicago, which opened in 1892. Johns Hopkins and Chicago started as graduate schools but soon added undergraduate colleges. The already existing graduate schools began to take research more seriously and other colleges added graduate schools. Through these moves the German universities made their impact on education here and high-level research was at least launched.

Although some advocates and entrepreneurs of research—Andrew Dickson White of Cornell, Charles W. Eliot of Harvard, and Gilman—believed that universities should undertake research, they also believed it should be subordinate to teaching. Imparting truth, White held, is more important than discovering it, and Eliot declared in 1869 that "the prime business of American professors in this generation must be regular and assiduous class teaching."

During his early years as president he was suspicious of professors who devoted much time to research because he feared that this activity interfered with their teaching. But a generation later the university leaders reversed the emphasis, and research took precedence over teaching. President William Rainey Harper of Chicago led this change, declaring: "The first obligation resting upon the individual members who comprise it [the university] is that of research and investigation." David Starr Jordan of Stanford declared, "The crowning function of a university is original research."

To defend its position the Harper-Jordan group asserted, in Jordan's words, that "investigation is the basis of all good instruction. No second-hand man was ever a great teacher and I very much doubt if any really great investigator was ever a poor teacher." Hundreds of professors and scores of university administrators took up the Jordan refrain that no one could be a good teacher unless he also did research. Harper not only made research the first obligation of professors; he also instituted the practice that promotion of the University of Chicago faculty "will depend more largely upon the results of their work as investigators than upon the efficiency of their teaching." Harper was also responsible for directing the graduate schools to concentrate on training all graduate students to be researchers. Breadth of knowledge was derogated and training for teaching was entirely ignored. The discovery of new facts or the recovery of forgotten facts became the supreme, prized goal.

Whether subordinated to teaching or esteemed more highly, research did take hold. One could say that it was effectively initiated when Johns Hopkins offered a professorship to the already famous British mathematician James Joseph Sylvester (who was refused a job in British universities because he was Jewish). Sylvester served from 1877 to 1883. He and William E. Story founded the first research journal in the United States, the *American Journal of Mathematics*, in 1878. This journal immediately became the outlet for the earliest significant research papers written by Americans; European contributions added to its prestige. The first volume contained several highly original papers by the American-educated mathematical astronomer George William Hill and in the volume of 1881 Benjamin Peirce published a paper he had written and circulated privately a decade earlier.

When the University of Chicago was organized it immediately hired outstanding research professors. In mathematics it appointed Eliakim Hastings Moore, an American who had studied in Germany, and Osker Bolza and Heinrich Maschke, both imported from Germany. This university became the first outstanding American research center, and it trained the first truly great American mathematicians, among them George David Birkhoff, Leonard Eugene Dickson, and Oswald Veblen.

The precedent set by Johns Hopkins, Stanford, and Chicago was soon followed by other universities, which now began to demand Ph.D.'s who could carry on research at their institutions. Some colleges also sought Ph.D.'s for the sake of the prestige gained by having such researchers on their faculties. However, the overall quality of the Ph.D. training was generally poor. The degree was aptly described by Jean-Paul Sartre as a reward for having a wealthy father and no opinions. The low quality of the Ph.D.'s, who were expected to become teachers as well as researchers, began to alarm prescient educators. In 1901 President Abbott Lawrence Lowell of Harvard said: "We are in danger of making the graduate school the easiest path for the good but docile scholar with little energy, independence or ambition. There is the danger of attracting an industrious mediocrity which will become later the teaching force in colleges and secondary schools." A few years later David Starr Jordan, despite his strong advocacy of research, deplored the lack of professors qualified to teach, which he attributed to the narrowness and triviality of the doctoral dissertation.

A more devastating and concerted attack was made by the distinguished Harvard philosopher, William James, in his essay of 1903, "The Ph.D. Octopus."* James was concerned that the rush for the Ph.D. crushed the true spirit of learning in the colleges. He objected to colleges and universities seeking Ph.D.'s as evidence to the world that they had stars on their faculties:

> Will anyone pretend for a moment that the doctor's degree is a guarantee that its possessor will be a success as a teacher? Notoriously his moral, social, and personal characteristics may utterly disqualify him from success in the classroom; and of these characteristics his doctor's examination is unable to take any account whatever.... In reality it is but a sham, a bauble, a dodge, whereby to decorate the catalogues of schools and colleges.

And in 1908, the distinguished American educator Abraham Flexner foresaw the greater evil to come, namely, that even though universities might improve the quality of the Ph.D. training, they would be sacrificing college teaching on the altar of research.

For better or worse, the emphasis on research grew stronger. To further it mathematicians decided to hold meetings and to support more journals. They founded the New York Mathematical Society in 1888, which became the American Mathematical Society in 1894. Initially the Society devoted itself to research and teaching. In the first decade or two of this century members of the Society—Eliakim H. Moore, Jacob W.A. Young, David Eugene Smith, Earle R.

*   See James's *Memories and Studies*, 1912.

Hedrick, George Bruce Halsted, and Florian Cajori, among others—did take an active interest in education, including secondary school teaching. They made sensible recommendations and seriously attempted improvements. But after a couple of decades the Society concentrated on research, whereupon another group of men founded the Mathematical Association of America in 1915 to cater specifically to undergraduate education. In 1921 still another group, concerned with secondary and primary school education, founded the National Council of Teachers of Mathematics.

More and more, research became the major interest of the universities, until in the large universities it gained favor over all other functions. Just how research and teaching might have fared had the United States continued its main reliance upon its own resources cannot be known. But unexpected developments altered the American scene. During the Hitler period many of the leading mathematicians of Germany, Italy, Hungary, and other European nations fled their countries; a large number of them came to the United States. When the Institute for Advanced Study was founded at Princeton in 1933 for postdoctoral research, three of the six mathematics professors chosen for its faculty were Albert Einstein, John von Neumann, and Hermann Weyl. The numerous refugees soon found places in American universities and added enormously to our mathematical strength. They also trained doctoral students for research, and the number and quality of Ph.D.'s increased significantly.

Though the acquisition of competent researchers was a boon to that activity, it contributed little to the teaching at the college level. The refugee mathematicians who came here during the Hitler terror were trained in the German universities, which were—until they lost their best professors—the strongest in the world. However, there was and is no undergraduate education in Germany. Students go directly from a *gymnasium* (high school) to a university, where they specialize in a subject, usually to obtain the Ph.D. degree. Training for research is the goal of the education (though students who do not complete the degree may pass an examination that qualifies them to be *gymnasium* teachers). Moreover, the professors, who are research specialists, do not feel obliged to be concerned with pedagogy, since student motivation, drive, and ability are presumed. Though the refugee mathematicians were intelligent and well intentioned they were not, by reason of background, able to apportion their efforts and knowledge among the diverse needs of the American universities, such as undergraduate education. The graduate schools were therefore turned still more in the direction of training researchers and became even less concerned with pedagogy.

While the graduate schools were absorbing the great academicians who had come to the United States, World War II broke out. The war was a battle of scientists. Faster ships and airplanes, radar for detection and to advance anti-aircraft gun control, improvements in range and accuracy of artillery, better navigational techniques for ships, planes, and submarines, and the development of the atomic bomb proved to be crucial and convinced the country that if it was to remain a major power, more mathematical and scientific research was needed. Hence, during and after World War II our government began to support research on a large scale. Research became a more and more valued and prestigious activity, and professors concentrated on this work. With the rising importance of research came the rising esteem for the researcher. The American professor, once a lowly figure among the elite, had now achieved high social rank.

Government money also made an essential difference. Billions of dollars to finance research were given to professors, who thereupon became sources of income to the universities. On paper all research grants and contracts cover only the actual cost of the research performed on behalf of some governmental agency or program. But in effect these grants cover far more. They support graduate students employed to aid in the research. As a consequence more graduate students can attend the university, and more tuition income is received. Professors working on contracts give up most of their teaching duties. In their place poorer paid, often young, instructors conduct most of the classes, though the name of a well-known professor still serves as a drawing card in attracting students. Laboratories and equipment needed to perform contract research are paid for by the government but are used for other research and for instructional purposes. Secretaries to professors, paid by contract money, and a generous overhead also make contract research very attractive. It does enable a university to enlarge its research and graduate programs, though— contrary to claims often made—it does not contribute to undergraduate education.

The effect was obvious. The universities, already inclined to favor research, began to compete intensively for research professors. Most of these professors and most graduate schools turned their attention to producing more research-ers. The Ph.D. programs were geared solely toward this end.

The attention that academic professors had given to undergraduate, secondary, and elementary school educa-tion during the first few decades of this century was all but abandoned. Nevertheless, there would no longer seem to be cause for the concern expressed by Lowell in 1901. The United States now has many strong graduate schools staffed by competent research professors, and research, judged at least by the volume of output, is flourishing. Mathematics research reinforces scientific strength, and scientifically and technologically the country has achieved pre-eminence. The researchers can certainly offer sound graduate training, and the knowledge imparted in this training should filter down to all levels of education. Wise professors, concerned not only with the extension of knowledge but also with its transmission, can devise ways of presenting their subjects that their students—the professors and teachers of the future— could use to good advantage.

The recently organized large schools or departments of education in the universities and colleges can provide the special instruction and advising that the elementary and secondary school teachers require. Insofar as mathematics in particular is concerned, although the American Mathematical Society has abandoned its earlier interest in teaching, the Mathematical Association of America and the National Council of Teachers of Mathematics have taken over that vital concern for the undergraduate and lower levels. On the face of things, the United States would seem to have reached, if not an ideal, then certainly a very reasonable and even rosy position at all levels of education.

Why, then, did Peter Landers face so many problems as a teacher and find himself unable to resolve them?

3

The Nature of Current Mathematical Research

Even victors are by victory undone.

John Dryden

We have traced the widespread rise of vigorous mathematical research in this country. We have also observed that the flourishing of research promises not only direct benefits but also indirect beneficial influence on all levels of education. However, the values that might accrue from research depend on the quality of research being done. Let us therefore look into the nature of current mathematical research.*

In mathematics, research has a very special meaning. Specifically, it calls for the creation of new results, that is, either new theorems or radically different and improved proofs of older results. Expository articles, critiques of trends in research, historical articles or books, good texts at any level, and pedagogical studies do not count. Thus, the

---

* Though we shall discuss mathematical research, many of its features, as earlier noted, apply to other academic disciplines as well.

criterion of research in mathematics differs considerably from what is accepted in, for example, a subject such as English. In this area, in addition to the creation of fiction, essays, poetry, or other literature, criticism, biographies that shed fresh light on important or even unimportant literary figures, histories of literature, and texts that may be primarily anthologies are considered original work. Perhaps this distinction between what should be accepted in the respective fields is wise, but let us see what it has led to in mathematics.

Because the United States entered the world of mathematical research several hundred years after the leading Western European countries had been devoting themselves to it, our mathematicians, in an endeavor to compete, undertook special directions and types of investigations.

One move was to enter the newer fields, such as the branch of geometry now called topology. The advantage of a new field for tyros in research is that very little background is needed and the best concepts and methodologies are only dimly perceived. Hence, because criteria for value are lacking, almost any contribution has potential significance. Publication is almost assured.

Of course, the ease with which one can proceed in a new field is somewhat deceptive. New fields generally arise out of deep and serious problems in older fields, and anyone who wants to do useful work must know much about these problems and grapple with them at length in order to secure significant leads. On the other hand, if all one is trying to do is prove theorems, then it is sufficient to start with almost any potentially relevant concept and see what can be proved about it. And if one gets a result that the other fellow didn't get, one may proceed to publish it.

The United States was not the only country that took such a course. After World War I, Poland was reconstituted as a nation and the Polish mathematicians undertook a concerted effort to build up mathematics in their country. They decided to concentrate on a narrow field, the branch of topology called point set theory. Why point set theory? Because at that time the subject was still new. One could therefore start from scratch, introduce some concepts, lay down some axioms, and then proceed to prove theorems. This example is offered not to malign Polish mathematicians. There were and are some very good men among them, and good men, even starting from very shallow beginnings, will make progress and produce fine work. What is significant is the deliberate and openly stated decision to start with point set theory because one did not have to know much mathematics to work in it.

Generalization is another direction of research that promises easy victories. Whereas the earlier Greek and European mathematicians were inclined to pursue specific problems in depth, in recent years many researchers have turned to generalizing previous results. Thus, while the earlier mathematicians studied individual curves and surfaces, many twentieth-century mathematicians prefer to study classes of curves—and the more general the class, the more prized any theorem about it. Beyond generalizing the study of curves, mathematicians have also carried most geometric studies to n-dimensions in place of two or three.

Some generalizations are useful. To learn how to solve the general second degree equation $ax^2 + bx + c = 0$, where a, b, and c can be any real numbers, immediately disposes of the problem of solving the millions of cases wherein a, b, and c are specific numbers.

But generalization for the sake of generalization can be a waste of time. A lover of generalization will too often lose sight of desirable goals and indulge in endless churning out of more and more useless theorems. However, those for whom publication is the chief concern are wise to generalize.

Hermann Weyl, one of the foremost mathematicians of this century, expressed in 1951 his contempt for pointless generalizations, asserting: "Our mathematics of the last few decades has wallowed in generalities and formalizations." Another authority, George Polya, in his *Mathematics and Plausible Reasoning*, supported this condemnation with the remark that shallow, cheap generalizations are more fashionable nowadays.

Mathematicians of recent years have also favored abstraction, which, though related to generalization, is a somewhat different tack. In the latter part of the nineteenth century mathematicians observed that many classes of objects—the positive and negative integers and zero; transformations, such as rotations of axes; hypernumbers, such as quaternions (which are extensions of complex numbers); and matrices—possess the same basic properties.

Let us use the integers to understand what these properties are. There is an operation, which in the case of the integers is ordinary addition. Under this operation the sum of two integers is an integer. For any three integers, $a + (b+c) = (a+b) + c$. There is an integer, 0, such that $a + 0 = 0 + a = a$. Finally, for each integer, a, for example, there is another integer, $-a$, such that $a + (-a) = -a + a = 0$. These properties are more or less obvious in the case of the positive and negative integers.

But if in place of the integers we now speak of a set of objects, which might be transformations, quaternions, or matrices, though the particular set is not specified; and of an operation, whose nature depends on the particular set of objects but is also not specified, we can state in abstract language that the elements of the set and operation possess the same four properties as those of the integers. The abstract formulation defines what is called technically a group. A group, then, is a concept that describes or subsumes the basic properties of many concrete mathematical collections and their respective operations under one abstract formulation. If one can prove, on the basis of the four properties of the abstract group, that additional properties necessarily hold, then these additional properties must hold for each of the concrete interpretations or representations of the group.

The concept of a group, very important for both mathematics and physics, is only one of dozens of abstract systems or structures—the latter is the fashionable word—and many mathematicians devote themselves to studying the properties of these structures. In fact, the study of structures is flourishing; the work done on groups alone fills many volumes.

Abstraction does have its values. One virtue, as already noted, is precisely that one can prove theorems about the abstract system and know at once that they apply to many concrete interpretations instead of having to prove them separately for each interpretation. Further, to abstract is to come down to essentials. Abstracting frees the mind from incidental features and forces it to concentrate on crucial ones. The selection of these truly fundamental ones is not a simple matter and calls for insight. Nevertheless, there can be shallow and useless abstractions as well as deep and powerful ones. The former are relatively easy to make, and one must distinguish this type of creation from that involved in solving a new and difficult problem—such as proving, as Newton did, that the path of each planet, moving under the gravitational attraction of the sun, is an ellipse or the far more difficult problem, which has still not been solved, of finding the paths of three bodies when each attracts the others under the force of gravitation. Unfortunately, many recent abstractions have been shallow.

Beyond the shallowness of some abstractions, there are

other negative features of all abstractions. Although unification through abstraction may be advantageous, mathematics pays in loss of resolution for the broadened abstract viewpoint. An abstraction omits concrete details that may be vital in the solution of specific problems. Thus, the manner of executing the processes of adding whole numbers, fractions, and irrational numbers is not contained in the group concept. The more abstract a concept is, the emptier it is. Put another way, the greater the extension, the less the intension.

Abstraction introduces other objectionable features. As a theory grows abstract it usually becomes more difficult to grasp because it uses a more specialized terminology, and it requires more abstruse and recondite concepts. Moreover, unrestrained and unbridled abstraction diverts attention from whole areas of application whose very investigation depends upon features that the abstract point of view rules out. Concentration on proofs about the abstraction becomes a full-time occupation, and contact with one or more of its interpretations can be lost. The abstraction can become an end in itself, with no attempt made to apply it to significant concrete situations. Thus, the abstraction becomes a new fragment of mathematics, and those fields that were to receive the benefits of unification and insight are no longer attended to by the unifiers.

Weyl spoke out against unrestrained abstraction, maintaining that "in the meantime our advance in this direction [abstraction] has been so uninhibited with so little concern for the growth of problematics in depth that many of us have begun to fear for the mathematical substance."

The inordinate attention given to the study of abstract structures caused another mathematician to warn, "Too many mathematicians are making frames and not enough are making pictures."

Another popular direction of research may be described roughly as axiomatics. To secure the foundations of their subject the late-nineteenth-century mathematicians turned to supplying axiomatic bases for various mathematical developments, such as the real number system, and to improving those systems of axioms where deficiencies had been discovered, notably in Euclidean geometry. Since there are dozens of branches of mathematics, there are dozens of systems of axioms. Quite a few of these contain ten, fifteen, or twenty axioms. The existence of such systems suggests many new problems. For instance, if a system contains fifteen axioms, is it possible to reduce the number and still deduce the same body of theorems? Given a system of axioms, what would be the effect of changing one or more of them? The classic and notable instance of this last-mentioned type of investigation is, of course, the change in the Euclidean parallel axiom and the resulting creation of hyperbolic non-Euclidean geometry. Changes in several of the axioms led to elliptic non-Euclidean geometry. Clearly, if a system contains as many as fifteen axioms, the changes that can be considered are numerous.

The investigation of the consequences of changing the Euclidean parallel axiom was indeed sagacious. By contrast present-day mathematicians, with little reason to do so, pursue all sorts of axiomatic investigations so that in the eyes of many practitioners, mathematics has become the science of axiomatics. The current activity in this area is enormous and overstressed. When axioms were believed to be self-evident truths about the constitution of the physical world, it was laudable to simplify them as much as possible so that their truth could be more apparent. But now that axioms are known to be rather arbitrary assumptions, the emphasis on deducing as much as possible from, say, a minimum number of axioms, which are often flagrantly artificial and chosen merely to reduce the number, is not warranted. The objective seems to be to produce more theorems per axiom, no matter how distorted and unnatural the axioms may be. Consequently, one finds long papers with tedious, boring, and ingenious but sterile material. Nevertheless, the popularity of axiomatics is readily understood. It does not call for the imaginative creation of new ideas. It is essentially a reordering of known results and offers many minor problems.

In his 1951 critique of current features of mathematical research, Weyl included axiomatics:

> One very conspicuous aspect of twentieth century mathematics is the enormously increased role which the axiomatic approach plays. Whereas the axiomatic method was formerly used merely for the purpose of elucidating the foundations on which we build, it has now become a tool for concrete mathematical research.... [However] without inventing new constructive processes no mathematician will get very far. It is perhaps proper to say that the strength of modern mathematics lies in the interaction between axiomatics and construction.

Still another questionable activity in modern axiomatics, derogatively termed "postulate piddling," involves the adoption of axioms merely to see what consequences can be derived. A prominent mathematician of our time, Rolf Nevanlinna, has cautioned: "The setting up of entirely *arbitrary* axiom systems as a starting point for logical research has never led to significant results.... The awareness of this truth seems to have been dulled in the last few decades, particularly among younger mathematicians."

Felix Klein, a leading German mathematician who was active from about 1870 to 1925, remarked that if a mathematician has no more ideas, he then pursues axiomatics. Another distinguished professor once remarked that when a mathematical subject is ready for axiomatization it is ready for burial and the axioms are its obituary.

The several directions research has taken point up the fact that there are soft and hard problems—or soft and hard research. In the days when the density of good mathematicians was high, soft problems were not often tackled. Moreover, nineteenth-century mathematicians, who were the first to grasp the advantage of abstract structures, faced a higher order of difficulty than present-day mathematicians face in that type of research. In recent times soft problems have been the ones most often tackled, and even if the proofs are complicated, the results may still be merely difficult trifles.

There is still another feature of mathematical research that affects seriously the interaction of research and teaching— the chief concern of this book. Whether it involves generalization, abstraction, or axiomatics or pursues some other direction, modern research is commonly acknowledged to be almost entirely pure—as opposed to applied. Pure research may be characterized as mathematics for mathematics' sake. That is, however the theme or problem is obtained, the reasons for undertaking it may be aesthetic interest, intellectual challenge, or sheer curiosity: "Let's see what we can prove." This is the motivation in axiomatics when a researcher rather arbitrarily decides to change an axiom just to find out what changes this entails in the resulting theorems. Applied mathematics, on the other hand, is concerned with problems raised by scientists, or with a theme that a researcher believes is potentially applicable.

There is no doubt that the problems of applied mathematics are more difficult. The branches of mathematics customarily associated with applied mathematics are now several hundred years old, and the giants of mathematics have worked in them. Anyone who wants to do something significant today in partial differential equations, for example, must have quite a background. And for the processes of idealization and model building in applied mathematics one must have intimate knowledge of the relevant physical field in order not to miss the essence of the phenomenon under study. (See also Chapter 7.)

Pure mathematics is more accessible for another reason. Whereas in applied work the problem is set by scientific needs and cannot be altered, the pure mathematician tackling problem A may, if unable to solve it, convert it to problem B, which could be A with more hypotheses or a related but actually different problem suggested by the work on A. He may end up solving problem B, or while working on it he may find unexpectedly that he can solve problem C. In any case he has a result and can publish it. In other words, the applied mathematician is required to climb a rugged, steep mountain, whereas the pure mathematician may attempt such a climb, but if he finds the going tough he can abandon it and settle for a walk up some nearby gentle hill.

Traditionally mathematics had been concerned with problems of science. But these, as we have noted, are far more difficult to solve. Only relatively few men today pursue them. The abandonment of tradition and of the rich source of problems has been justified by a new doctrine: Mathematics is independent of science, and mathematicians are free to investigate any problems that appeal to them. The research done today, so it is claimed, will be useful ten, fifty, and one hundred years from now. To support this contention the purists distort history and point to alleged examples of such happenings. But a correct reading of history belies the contention. Practically all of the major branches of mathematics were developed to solve scientific problems, and the few that today are pursued for aesthetic satisfaction were originally motivated by real problems. For example, the theory of numbers, if one dates its beginning with the Pythagoreans, was undertaken for the study of nature. Nevertheless, the break from science has widened sharply since about 1900, and today most mathematicians no longer know any science or even care whether their work will ever have any bearing on real problems.

Marshall Stone, formerly a professor at Yale, Harvard and Chicago, in an article "Mathematics and the Future of Science" (1957) admits that generality and abstraction—pure mathematics generally—are the chief features of modern mathematics in our country. The best applied mathematics, he concedes, is done by physicists, chemists, and biologists. He might well have added that mathematics developed in a vacuum proves to be vacuous.

Quite a different feature of modern research is specialization. The worldwide spread of scientific and technological pursuits has made it impossible for any individual to keep pace with a broad spectrum, and the desire to avoid being beaten to results by an ever-increasing number of competitors, and thus lose the fruit of months of activity, has almost forced mathematicians to seek out corners of their own. Mathematics is now fragmented into over a thousand specialties, and the specialties multiply faster than amoebas. The many disciplines have become autonomous, each featuring its own terminology and methodology. A general meeting of mathematicians resembles the populace of Babel after God had confounded their efforts. Pure mathematicians are unable to communicate with applied mathematicians, specialists with other specialists, mathematicians with teachers, and mathematicians with scientists. It is almost a certainty that if any two mathematicians were chosen at random and shut up in a room they would be so unintelligible to one another as to be reduced to talking about the weather. Consequently, general meetings are now far less numerous than colloquia and conferences on particular topics.

Illustrations of the narrowness of modern research are so abundant that almost any article in any journal can serve as an example. Let us note one or two simple ones. One article treats powerful integers. An integer is powerful if whenever it is divisible by a prime p it is divisible by p^2. Several papers on this less-than-enthralling theme have already appeared and more are sure to follow. Would that the papers be more powerful than the concept. Still another theme deals with admirable numbers. The Pythagoreans of the sixth century B.C. had introduced the concept of a perfect number. A number is perfect if it equals the sum of its divisors (other than the number itself). Thus $6 = 1 + 2 + 3$. If the sum of the divisors exceeds the number, the number is called abundant. Thus, 12 is abundant because the sum 1+2+3+4+6 is 16. One can, however, ask about the *algebraic* sum of the divisors; that is, one can consider adding and subtracting divisors. Thus $12 = 1+3+4+6-2$. Numbers that are the algebraic sum of their divisors are called admirable. One can now seek admirable numbers and establish properties about them, which no doubt are equally admirable. In this same vein are a superabundance of theorems on superabundant numbers.

These very trivial examples are, of course, chosen merely because they can be presented quickly to illustrate the narrowness and pointlessness of much modern-day research. Just as everyone who daubs paint on canvas does not necessarily create art, so words and symbols are not necessarily mathematics.

Specialization began to be common in the late nineteenth century. Now most mathematicians work only in small corners of mathematics, and quite naturally each rates the importance of his area above all others. His publications are no longer for a large public but for a few colleagues. The articles no longer contain any indication of a connection with the larger problems of mathematics, are hardly accessible to many mathematicians, and are certainly not palatable to a large circle. Mathematical research today is spread over so many specialties that what was once incorrectly said of the theory of relativity does apply to the research: Any one topic is understood by no more than a dozen people in the world.

Each mathematician today seeks to isolate himself in a domain that he can work for himself and resents others who might infringe on his domain and secure results that might rob him of the fruits of his work. Even Norbert Wiener, one of the great mathematicians of recent times, admitted that he "did not like to watch the literature day by day in order to be sure that neither Banach nor one of his Polish followers had published some important result before me." And the late Jacques Hadamard, the dean of French mathematicians until his death at the age of ninety-eight in 1963, said, "After having undertaken a certain set of problems and seeing that several other authors had begun to follow the same line, I would drop it and investigate something else."

There is a way of joining the crowd and yet keeping aloof from the hurly-burly. A favorite device is to introduce some new concept and develop endless theorems whose significance is, to say the least, questionable. The creator of such contrived material may even train doctoral students who, young in the ways and judgment of mathematics, may really believe in the worth of the material and so spread the name of the master.

Most of those working in specialties no longer know why the class of problems they are working on was originally proposed and what larger goals their work is supposed to aim at. The modern topologist may not know Riemann's and Poincaré's work. The modern worker in Lie algebras is not likely to know what purpose Lie algebras serve. Of course, these specialists are putting the cart before the horse. The limited problems should contribute to and illuminate the area in which they lie. But the specialists would seem to be taking the position that the major areas exist in order to provide problems on which to exercise their ingenuity. Nor do they recognize that specialization promotes one's degeneration into a narrow, uncultured person, a craftsman but nothing more. The specialist becomes what José Ortega y Gasset called a "learned ignoramus."

As the process of subdivision progresses, specialized research makes less and less provision for synthesis, for pulling strands together, for asking the basic, overriding questions, for stepping back from the easel and looking at the whole picture. Indeed, specialized research does not concern itself with synthesis. Though it may foster localized competence, it may simultaneously rationalize, and even glorify, general ignorance and deliberate unconcern for those questions that transcend the narrow bounds of specialism. Yet these questions are the ones that make sense of the whole enterprise.

Rampant specialization turns out to be a misfortune for the specialized pursuits themselves, although it seems to arise through concern for their exclusive needs. One obvious reason is that specialization encourages uninhibited intellectual inbreeding and it is a law, not only of human genetics, that inbreeding increases the incidence of undesirable characteristics. Furthermore, the process of unlimited specialization tends to bar a subject from the interest and participation of anyone outside, even when the outsider could make an essential contribution toward maintaining relevance in the questions asked and the methods used to pursue them. It also dims awareness of the fact that the pursuit of truth is indivisible, that all creative scholars, writers, and artists are ultimately engaged in one great common enterprise—the search for truth. In other words, specialization curtails the basic commitment of the scholar.

The evils of specialization have been noted by many wise men. In his history of nineteenth-century mathematics (1925), Felix Klein said that academic mathematicians grow up in company with others like trees in a woods, which must remain narrow and grow straight up in order even to exist and reach some of the light and air.

Weyl said in 1951, "Whereas physics in its development since the turn of the century resembles a mighty stream rushing in one direction, mathematics is more like the Nile delta, its waters fanning out in all directions." In the preface to his book, *The Classical Groups* (2nd ed., 1946), he expressed concern about too much specialization in mathematics: "My experience has seemed to indicate that to meet the danger of a too thorough specialization and technicalization of mathematical research is of particular importance for mathematics in America."

David Hilbert, the greatest mathematician of this century, was also concerned about specialization. He wrote:

"The question is forced upon us whether mathematics is once to face what other sciences have long ago experienced, namely, to fall apart into subdivisions whose representatives are hardly able to understand each other and whose connections for this reason will become ever looser. I neither believe nor wish this to happen; the science of mathematics as I see it is an indivisible whole, an organism whose ability to survive rests on the connection between its parts.

The trend to specialization has already caused mathematics departments to split into four or more departments— pure mathematics, applied mathematics, statistics and probability (with antagonism between the two groups in this area portending a future split), and computer science. Communication among these departments is, of course, almost nonexistent, and competition for money, faculty, and students is keen.

Clearly, the evils of specialization lead to inferior work. Specialists define their own area of interest and, as we have already noted, choose areas in which they can avoid competition and the larger, more vital problems. Publication is the goal, and whatever results can be published are published. Ortega y Gasset remarked in his *The Revolt of the Masses* that specialization provides what the biologist would call ecological niches for mediocre minds.

Since specialization is the order of the day, why not journals for specialists? These are now by far the most numerous, and specialists read only the journals in their own areas, thus precluding even their awareness of anything outside their specialty. There are few journals that cover— and none that unify—developments in several fields, to say nothing about all fields of mathematics.

Mathematical research has always suffered from another evil: faddism. Like all human beings mathematicians yield to their personal enthusiasms or are ensnared by the fashions of their times. The directions of research are often determined by mathematicians with prestige and power who themselves are subject to whims or the search for novelty. In the nineteenth century, for example, the study of subjects such as elliptic functions, projective geometry, algebraic invariants, and special properties of higher-degree curves was carried to extremes. Most of this work, considered remarkable in its time, would be considered insignificant today and has left almost no trace in the body of mathematics.

It is no criticism of mathematicians that an area of research pursued vigorously for a time should prove unimportant in the long run. Mathematicians must use judgment as to what may be worthwhile, and even the wisest can make mistakes. Research is a gamble and one can't be sure that the work will pay off. However, faddism tends to carry a subject beyond any promise of significance.

Fads flourish today because usefulness to science is no longer a standard, and the standard of beauty is purely subjective. The most pointed criticism of faddism was made by Oscar Wilde: A fad is the fantastic which for the moment has become universal.

Another evil of faddism is that possibly valuable but nonfashionable ideas are disparaged. Hence, brilliant work is often neglected, though it is sometimes belatedly and often posthumously recognized. The classic example is found in the work of Gauss. Gauss, though already acknowledged as great when still a young man, feared to publish his work on non-Euclidean geometry because he would have been condemned by his fellow mathematicians or, as he put it, because he feared the clamor of the Boeotians, a reference to a dull-witted ancient Greek tribe. Fortunately, Gauss's work on non-Euclidean geometry was found among his papers after his death. By that time his reputation was so great that his ideas were accorded the utmost respect.

Researchers who place high value on their work should be obliged to read a somewhat detailed history of mathematics, a subject most mathematicians do not know. They would be amazed to find how much that was regarded as vital and central in the past has been dropped so completely that even the names of those activities or branches are no longer known. Though the lesson of history is rarely learned, fads do not, fortunately, dominate the directions of research for long. What individuals create is destined to live only insofar as it is related to the evolutionary development of mathematics and proves fruitful in its consequences.

One additional source of research papers of dubious value should be mentioned—Ph.D. theses and their offspring. The students, beginners in research, cannot tackle a major problem; what they do tackle is not only suggested by a professor but is performed with his help. The results are generally minor and, in fact, usually the professor can see in advance how to solve the problem. If he could not, he might worry about whether he is assigning too difficult a problem to a beginner.

The new Ph.D. is, in today's world, forced to produce low quality research. If he enters or seeks to enter the university field, where researchers are now more sought after, he is under pressure to publish. Under these conditions what will he publish? He is at a stage in life where he is really not prepared to publish a paper of quality. Typically, his one experience in research was his doctoral thesis, in which he was guided by a professor and gained only enough knowledge to produce an acceptable thesis. Hence, all he actually is prepared to do is add tidbits to his thesis. But he cannot afford to be deterred by the knowledge that his publications may be insignificant. Some publication is better than none. Were he to try to solve a deep problem requiring extensive background and several years to complete, with the danger of failure all the greater, he would have nothing to show for quite some time, if ever. Hence, he must tackle and publish what can be done readily, even if the solution is labored and the result pointless.

The results of pressure on faculty and young scholars to publish, the natural expansion of research in our scientifically oriented culture, the entry of the Soviet Union, China, and Japan during this century into the group of countries leading in research, and the expansion of Ph.D. training to meet the needs of universities and colleges (which in recent years has meant 750 to 1,000 mathematics Ph.D.'s per year in the United States alone) are reflected in the volume of publication. There are now over a thousand journals devoted wholly or partially to mathematical research. About five hundred are devoted solely to mathematics, and new ones are appearing almost weekly. Summaries of the articles are published in *Mathematical Reviews*, which does not cover all articles and in fact neglects pedagogy and much applied mathematics. In 1970 there were 16,570 reviews; in 1973, about 20,000. If all applied mathematics had been covered there would have been about 40,000 reviews in 1973. The expansion of publications has been going on at the rate of 5 percent annually. In the period 1955 to 1970 the volume of publication equaled the volume in all of the rest of recorded history. The published papers are about one-fourth of those submitted to journals. Hence, one can see how much effort is put forth by faculty to climb the ladder of research.

To help mathematicians keep track of what has been published, secondary and even tertiary aids, such as indices and lists of titles, have been developed. There is an *Author Index of Mathematical Reviews*, which lists by author and subject the summaries published in *Mathematical Reviews*. For the years 1940 to 1959 *The Index* has 2,207 pages. For 1965 to 1972 it has 3,032 pages and 127,000 items, whereas the *Indexes* of the previous twenty-five years, 1940-1964, covered 156,000 items in all. There is also a journal, *Contents of Contemporary Mathematical Journals* (biweekly), that offers an index classified by subject of all current papers and books in mathematics. About 1,200 journals are covered and these do not include some in applied mathematics. We may await momentarily an index of the *Contents* and an index of all indices.

The volume of publication has evoked critical comments

from prominent mathematicians. One of them, Peter Hilton, has written, "...we are all agreed that far too many papers are being written and published. We are turning into a community of writers who do not read simply because we have no time to do so. It is a terrifying thought that if we were to spend eighteen hours a day reading new mathematics we would have substantially more to read at the end than at the beginning." In addition to zero population growth this country should aim for zero publication growth.

It was generally agreed in the 1930s, when the pace of research was much slower, that nine out of ten papers had little to say and had no impact on mathematics. Some significant quantitative information was supplied by Kenneth O. May, a professor at the University of Toronto, who studied the nearly two thousand publications from the seventeenth century to 1920 on the limited topic of determinants. He presents the following data:

New ideas and results	234	14%
Duplication (beyond independent simultaneous publication)	350	21%
Texts and education	266	15%
Applications of results	208	12%
Systematization and history	199	12%
Trivia	737	43%
Totals including overlap	1994	117%

The explanation of the 117 percent is that some papers fell into two or more categories; actually there were 1,707 separate papers. Professor May estimated that the significant information about determinants, including the main historical accounts, is contained in less than 10 percent of the papers. He also mentions that in 1851 there were ten duplications of a paper published in a leading journal.

Today, with far more papers published and far less

concern for the significance of the research, one might estimate that no more than 5 percent of the publications offer new material. The duplication is endless. Some of it is noted in *Mathematical Reviews. The American Mathematical Monthly* occasionally reports duplications and errors and even cites instances of purportedly new research material that has already appeared in texts.* This is not to say that all of the other 95 percent are wasted. A few have educational value. Nevertheless, the journals are filled with papers of flea-sized significance, and these pollute the intellectual world as noxiously as the automobile pollutes the air we breathe.

Authors deliberately publish minor variants of older research or repeat older results in new terminology. Unfortunately, the introduction of new terminology is a never-ending game, and a translation of old material can pass undetected, just as a French paper must be accepted as new by one who can't determine whether it has appeared in German. One famous nineteenth-century German mathematician did simply translate English papers into German and publish them as his own. Some researchers take one reasonably coherent paper and break it up into three or four smaller ones. This stratagem permits much repetition, thus resulting in more published pages and giving the impression of a teeming mind.

The profusion of articles and the ever-increasing number of journals make it impossible for even the specialist to read what is published in his own area. Hence, though he may pretend to know what has been done, he actually ignores the literature except for the few papers that he happens to know bear directly on his immediate goal—publication of his own paper. Months or years later some observer may note a duplication and call attention to it.

*   See, for example, the issue of December 1976, pp. 798-801.

Apart from the expense involved, the flood of papers seriously hampers research. A conscientious researcher will try to keep abreast of what is being done in his area, partly to utilize the results already obtained and partly to avoid duplication. He must then wade through a vast number of papers at the expense of considerable time and effort, and at that he will not cover all the relevant literature.

The problem of keeping abreast of the literature had already begun to bother Christian Huygens in the late seventeenth century. In 1670 he complained, "...it is necessary to bear in mind that mathematicians will never have enough time to read all the discoveries in geometry (a quantity that is increasing from day to day and seems likely in this scientific age to develop to enormous proportions) if they continue to be presented in a rigorous form according to the manner of the ancients." Leibniz at the end of the seventeenth century deplored "this horrible mass of writing which continually increases" and which can only "drive away from the science those who might be tempted to indulge in it."

In mathematics, where the newness of a result should be readily recognized and the difficulties overcome in proof readily apparent, it would seem that papers would be easily and accurately evaluated. Most journals do send manuscripts to referees before accepting them. But the good mathematicians who might serve as referees are so busy doing their own research, and the volume of publication they must follow is so enormous, that most do nothing about judging work in their own specialty, to say nothing of other areas of mathematics. Moreover, most papers are so sparse in explanation that their correctness is hard to judge.

The narrowness of mathematicians also renders them unfit to discriminate between what is fundamental and what is trivial, between basic insight and mere technical byplay. For interdisciplinary papers it is almost impossible to find competent referees. Personal factors also intervene. Individuals favor friends and discriminate against rivals.

The state of refereeing is revealed by the reactions to a recent decision of the American Mathematical Society. Up to 1975 all papers submitted for publication in the several journals supported by the Society were sent to referees with the names and affiliations of the authors recorded on the papers. The Society decided to try, for one of its journals, blind refereeing, that is, submitting the paper to the referee without the name and affiliation of the author. The protests of referees and even of two of the associate editors of that journal were vehement. They pointed to the thanklessness of the work, the difficulty in finding competent referees, and the problem of judging the correctness and worth of a paper. In the ensuing debate, partly through published letters, the opponents of blind refereeing admitted that the name and affiliation of the author helped immensely in the refereeing process. What these opponents were really saying is that they were not judging papers on their merits but were relying on the reputation of the author and his institutional affiliation to aid in determining the correctness and value of his work. If one may judge by the protests, many referees used no more than this information to make their decisions. This debate brought into the open all the weaknesses of the refereeing process.

Moreover, today many papers are published without judgment by referees. There are countless symposia each year, and the papers read there are published automatically in the proceedings. Some universities produce their own journals, in which faculty members can publish at will. Publication in the *Proceedings of the National Academy of Sciences* is automatic not only for members but also for nonmembers whose papers are submitted through a member. The extent of the Academy's publications may be judged by the fact that in 1970 the editors decided to restrict each member to no more than ten papers per year.

The present situation contrasts sharply with what prevailed in the seventeenth, eighteenth, and nineteenth centuries. Of course, there were fewer publications. But papers were sent to referees who were not only distinguished mathematicians but also broad scholars. Even then there were slips in both acceptance and rejection. R.J. Strutt, the son of one of the greatest mathematical physicists, Lord Rayleigh, relates in his life of his father that a paper by Lord Rayleigh that did not have his name on it was submitted to the British Association for the Advancement of Science and was rejected as the work of one of those curious persons called paradoxers. However, when the authorship was discovered, the paper was judged to have merit. Nevertheless, on the whole the refereeing of earlier times was competent and critical. Moreover, the editors took pride in the quality of the work published in their journals and were anxious to maintain excellent reputations. They therefore took pains to secure competent criticism of articles submitted. It is also relevant that usefulness to science served as the major standard by which most papers were judged.

Actually, what is major or minor in research can be very difficult to determine. François Vieta, who first taught us to use letters to stand for a class of numbers, as in $ax^2 + bx + c = 0$, an idea that now seems trivial but was not advanced until after two thousand years of first-class mathematics had been created, gave mathematics the basis for all proof in algebra and analysis. Surely this idea was as valuable as any major result of Newton.

The assertion that quality of research is difficult to judge may seem to contradict our earlier assertion that most papers have little, if any, value. The worth of a few papers—for example, those that solve a long-standing problem that had baffled great minds—is certainly great. In other cases the authors state why the results they have obtained are important, so that the work can be more readily judged. When Vieta introduced letters for classes of numbers he stated that he could now make the distinction between numerical algebra and a science of algebra (to use modern terminology). Perhaps many a seemingly worthless paper has merit, but if that merit is not apparent to knowledgeable mathematicians, only an adverse judgment is in order.

Some sociologists of science are trying to measure the quality of research papers by the number of times a given paper is cited by later papers. Toward this end they invented and use the *Science Citation Index*. But this measure is almost childish. Very good papers are often soon superseded by ones that advance the subject still further. Even when the advances are minor, the later papers will surely be the ones cited. A fad will be cited many times over a period of years. Many young researchers cite their professors, even at the expense of the true creator, in order to curry favor. Accepting citations, then, would seem to require first measuring the honesty of scientists.

Modern mathematical research seems impressive. There is a vast and growing structure. Recent work has delineated more sharply the nature of the older subjects and has pointed the way to almost endless paths of new developments. Abstractions and generalizations have linked apparently unrelated subjects, giving mathematics some measure of unity, and have put some difficult classical theorems in a new setting where they become more natural and meaning-ful, at least to a trained mathematician. Mathematics now has a more qualitative character, in contrast to the manipulative and quantitative character of much of classical mathematics. Many new subjects have been created; and areas of older subjects that no longer seem significant have been discarded. We no longer learn all 467 theorems in Euclid's *Elements* or all 487 theorems in Apollonius' *Conic Sections.*

But a critical look produces dismay. The proliferation of new themes, generalization, abstraction, axiomatics, and specialization may yield easy successes, but they divert attention from more concrete and difficult problems concerned with ideas of substance. Abstractions and specialties abandon reality to enter clouds of thin and diffused themes. An overweeningly arrogant antipathy to papers that do not follow the modern fashions also encourages less valuable activity.

Mathematicians today care less and less about why mathematics should be created and pursued. They pay far less attention to what is worth knowing or what benefits society; nor do they question why society should support them. One of the most disturbing facts about current research is that graduate students, young Ph.D.'s, and even many established mathematicians no longer ask, Why should I undertake this particular investigation? Any inquiry that promises to produce answers and publication is regarded as worthwhile. No commendable purpose need be served except, perhaps, to advance the career of the researcher. A problem is a problem is a problem, and that suffices. Though criticism is rarely voiced, one past president of the American Mathematical Society and the Mathematical Association of America did have the courage to deprecate much modern research.

No doubt much worthless research is done in all academic fields. But remoteness and pointlessness are far more prevalent in mathematics. The reason stems from the nature of the subject, especially as it is currently pursued.

Mathematics deals not with reality but with limited abstractions. In past centuries these did come largely from real situations, and the prime motivation for the mathematics was to learn more about physical reality. It was recognized that the pursuit of well-chosen problems in mathematics proper must directly or indirectly pay dividends in scientific work, and mathematicians were obliged to keep at least one eye on the real world. But today mathematicians know better what to do than why to do it. The pointlessness of much current research is evident in the very introductions to papers. Students and professors seeking themes for investigation scan the publications and tag onto them. Many a paper begins with the statement, "Mr. X has given the following result.... We shall generalize it," or, "Mr. X has considered the following question.... A related question is...." There may be no point to either the generalization or the related question. Another common introduction states, "It is natural to ask...."; a most unnatural and far-fetched question follows. The consequence is a wide variety of worthless papers.

Mathematical research is also becoming highly professionalized in the worst sense of that term. Research performed voluntarily and sincerely by devoted souls, research as a relish of knowledge, is to be welcomed even if the results are minor. But hothouse-grown research, which crowds the journals and promotes only promotion, is a drag on science. Intellectual curiosity and the challenge of problems may still provide some motivation, but publication, status, prizes, and awards such as election to the National Academy of Sciences are the goals, no matter how attained. Deep problems that call for the acquisition of considerable background, years of effort, and the risk of failure are shunted aside in favor of artificial ones that can be readily tackled and almost as readily solved.

This indictment of current research may surprise many people. Surely mathematicians are men of intellect and would not write poor or worthless papers. But the quality of the intellects engaged in research runs the gamut from poor to excellent. Francis Bacon in his *Novum organum* (1620) sought to mechanize research and was rightly and severely criticized. His own contemporary Galileo demonstrated through his work the extent to which originality and serendipity must enter. Bacon may indeed have oversimplified the task of research, but his expectation that anyone can do it, even "men of little wit," is not far from what happens today in mathematical research.

Professor Clifford E. Truesdell, an authority in several applied fields and a man of vast knowledge, has had the courage to speak caustically. In his *Six Lectures on Natural Philosophy* he says:

> Just as the university has changed from a center of learning to a social experience for the masses, so research, which began as a vocation and became a profession, has sunk to a trade if not a racket. We cannot fight the social university and mass-produced research. Both are useful—useful by definition, since they are paid, if badly.... The politician, the lawyer, the physician, the general, the university official are all modest men, more modest than most mathematicians. ...Research has been overdone. By social command turning every science teacher into a science-making machine, we forget the reason why research is done in the first place. Research is not, in itself, a state of beatitude; research aims to discover something worth knowing. With admirable liberalism, the social university has declared that every question any employee might ask is by definition a fit object of academic research; valorously defending its members against attacks from the unsympathetic outside, it frees them from any obligation to intellectual discipline....

Though each mathematician must be free to pursue the research he prefers, he does have the responsibility to produce potentially applicable papers or papers that offer high aesthetic quality, novelty of method, freshness of outlook, or at least the suggestion of a fruitful direction of research. But far too many mathematicians take advantage of the facts that potential use is difficult to judge and aesthetic quality is a matter of taste. Hence, the good is swamped by the bad. Of course, as in the past, history will decide what is of lasting value. It is the deserved fate of inferiors to fall into oblivion. But the temporary profusion of ideas they introduce constitute today a hampering and almost insuperable obstacle to real progress.

Whether or not current research will prove more hindrance than help to the advancement of mathematics is, however, not our primary concern. We undertook to survey the nature of this research because its maturation seemed to promise improvements at all levels of our educational system. Let us, then, look into the relationship of research and teaching.

4

The Conflict Between Research and Teaching

What a blessed place this would be if there were no undergraduates! ... No waste of good brains in cramming bad ones.

Leslie Stephen

The major role of the universities is to carry on two functions—research and teaching. However, money and facilities are limited. The universities solve this problem by appointing research professors. Such people, they assert, advance knowledge and improve all levels of education directly and indirectly. More specifically they maintain that researchers are *ipso facto* good teachers. They also assert the converse: To be a good teacher one must be a good researcher. Hence appointment, promotion, tenure, and salary are based entirely on status in research.

Although there is justification for the insistence that professors who train future researchers be capable in research, for most of the teaching that the universities are, or should be, offering, the research professor is useless. There is in mathematics, and almost as surely in other disciplines, a direct conflict between teaching and research.

One argument advanced in favor of the research man is that he has superior knowledge. But what knowledge does he possess? He is almost sure to be a specialist. The specialist is like a miner who never surveys a landscape but who digs a barely passable tunnel into a mine to research a small lode of gold that may in fact prove to be tin. Such research narrows rather than broadens. What the professor does in his research has little if any bearing on what he has to teach at the undergraduate and even beginning graduate levels. A specialist in some corner of abstract algebra may know nothing about non-Euclidean geometry. In fact, the creative researcher is most likely to be no more than a proficient but limited technician in one minuscule area. The irrelevance of specialized knowledge for teaching is not peculiar to mathematics. A professor of English who has specialized in philology or in the love life of Madame Bovary is not by virtue of this research equipped to teach elementary composition or a survey of literature.

Not all mathematical research is specialized. For example, some work on abstractions and generalizations is rather broad. But, apart from the worthlessness of much of this material (see Chapter 3), this work is far too sophisticated for most of the teaching professors are called upon to do. Abstractions and generalizations are meaningful only to students who already have a considerable knowledge of more concrete mathematics. Unfortunately, researchers are prone to plunge at once into abstractions and generalizations because these are dear to their hearts.

The research professor's knowledge is of questionable value for another reason. Most mathematical research today is "pure"; it has no relevance to problems arising in the real world. But most of the students who take mathematics, about 95 percent, plan to be physical scientists, engineers, economists, actuaries, statisticians, and primary and secondary school teachers. The typical researcher's knowledge is useless to such students.

Beyond all the previous reasons for devaluating the knowledge research professors possess, there is an additional one that applies specifically to mathematics education. Mathematical knowledge is cumulative. Just as algebra depends on arithmetic, the calculus on algebra, differential equations on the calculus, and so on, so the researcher's knowledge is accessible only to those who have considerable prerequisite knowledge in his area of research.

A common argument often advanced in favor of the researcher is that he evidently loves his subject and will communicate his enthusiasm to his students. One must recognize, however, that research today is—as Professor Truesdell, whom we have already cited, and Blaise Pascal much earlier, called it—far more a trade than a profession, and many people engage in it merely as one way to make a living. Indeed, since the universities stress research as the road to personal advancement, many professors are pressed into research much as draftees are pressed into the army whether or not they are willing to fight. Research is a highly competitive activity, not a labor of love. The pressure to publish is exemplified by the message one chairman sent to the members of his department: "A theorem a day means promotion and pay."

Even the enthusiastic professor who is deeply engrossed in his research almost unconsciously assumes that his mission is to stir up interest in his specialty, and accordingly rushes his students toward it at a pace they cannot maintain and at the expense of far more valuable and basic material. He seeks to train specialists even though this training is useless to almost all of his students.

Moreover, since prominence and visibility in his field is demanded, the professor undertakes activities that enable him to shine. He will publicize his results at countless conferences and professional meetings. He will pursue grants and consulting arrangements that, beyond being profitable, aid his prestige. To attract attention he will seek office in a professional society. All these activities enhance his commercial value and further his personal advancement. What about obligations to students? This is the nethermost consideration. Even loyalty to his institution is not respected. Professors negotiate their own research grants and take them along when they transfer to another institution. In brief, they become denizens of the marketplace. Professors have no more conscience about obligations than has any other group chosen at random.

The layman tends to think of research as a genteel, leisurely activity, conducted with an open, disinterested attitude and devoted to the pursuit of truth. Actually it is a fierce battle for survival. Since priority in publication means everything, a researcher must push his work as fast as he can lest someone else beat him to it. Lord Ernest Rutherford, one of the greatest of British physicists, wrote in 1902, "I have to keep going, as there are always people on my track. I have to publish my present work as rapidly as possible in order to keep in the race. . . ." And the attempt to keep up with the literature in his field will absorb a considerable amount of a researcher's time and energy. The intensity of the competition even obliges researchers to adopt invidious practices more commonly attributed to the business world, such as refraining from telling others what they are trying to do or what partial success they may already have. For the sake of an immediate publication, a researcher may feel compelled to publish shoddy results, even though he may be quite sure that further efforts will produce a far better paper. Research requires an all-out effort and forces neglect or perfunctoriness in teaching.

The universities' insistence on research imposes a special hardship on young Ph.D.'s They have just emerged from indoctrination in the purity of mathematics and from the dark recesses of some specialty they have pursued for two or three years. The doctorate conferred upon them is not the certification of a teacher but the official stamp of cultural deprivation. They then receive a three-year appointment during which they are expected to demonstrate their brilliance through publication. But this is just the period when most begin to teach; they must spend time in the preparation of courses that are new to them and in learning much about the art of teaching. After two years of service it is customary to notify these teachers whether their appointments will be renewed. Actually, not even two years can be used to demonstrate their research strength; the journals are overcrowded and many months must elapse before a paper is refereed and even accepted for publication. Since up to the time of their appointment these young teachers did only one piece of research, the thesis, and that was done under the guidance of a professor, they face a formidable task. Even if good teaching is regarded with some favor in their institutions, tyros in teaching certainly cannot expect to shine in this capacity. There is but one human response to such a conflict. It is publish, and perish the students.

The folly of expecting young Ph.D.'s to justify their claim to appointment or tenure through publication is evidenced by the case of Albert Einstein. After obtaining a diploma at the Zurich Polytechnic Institute, he sought merely an assistantship in a German university in order to obtain a doctor's degree. But at that time he had no publication to demonstrate his capacity for research. He therefore took a job as an examiner in the Swiss patent office at Berne. Several years later, in 1905, he published three brilliant papers, including the one on the special theory of relativity. Six more years elapsed before he obtained his first university position at the German university in Prague.

The requirement of research from young Ph.D.'s was ridiculed recently by a professor of philosophy at an Ivy League institution: "Most of the great philosophers could not have gotten tenure at Blank University; they published their important work after they were forty. Kant didn't write anything for twelve years—he would have been out on his butt."

That young men and women are expected, however unreasonably, to show research strength is readily confirmed merely by glancing at employment advertisements in the professional journals. One prestigious university advertised for an assistant professor who would already have published work of high quality and would be expected to continue to be active in research. The applicant was also to present evidence of proved effectiveness in teaching at undergraduate and graduate levels. All this was required from a young man or woman who, if applying for an assistant professorship, would be a Ph.D. of, at most, two or three years' standing.

A not uncommon advertisement reads essentially as follows: Wanted: Young Ph.D.'s, ranks of assistant professor and higher. Must have proven research ability. Vita must include publications. Apply to Prof. I.M. Noble, Superior Community College, Worthmore, Nevada. (Salaries readily augmented. Short drive to Las Vegas.)

One would think that department chairmen, who can be in close contact with their young Ph.D.'s, would be able to judge potential without requiring publication. But the ability to discern talent is rarer than ability to do research, and many chairmen do not possess it. They often consult a colleague, who may be equally incapable of judging, or an outside specialist who may be biased. Moreover, young people are not likely to be known outside a limited academic circle.

The conflict between teaching and research becomes even more apparent when one considers the demands of teaching. Perhaps the first consideration is breadth of knowledge. Commerce students, prospective elementary and secondary school teachers, and social science students are now consumers of mathematics. Engineering and physical science students also need subject matter different from that for pure mathematicians. Just as it would be folly to require prospective dentists to learn Anglo-Saxon, so it is folly to require that prospective engineers learn the rigorous foundations of mathematics. Even on the graduate level most students are not preparing for mathematical research. The professor should know what material is important for them and fashion suitable courses. But the research mathematician cannot devote time to such matters and so teaches the material with which he is at ease.

What else does good teaching demand? Certainly it calls for knowing what backgrounds students bring to the class. Mathematics, at the lower levels especially, is a sequential subject, and a course must start where the prior education left off. To know what freshmen bring to college calls for knowing what the high schools teach. This is especially true today because the high schools have been changing curricula under the influence of new movements. But research professors will not trouble to find out what is being taught.

A good teacher gets to know his pupils. Some can be asked to tackle the more difficult problems or do additional work. The poorer ones should be called upon only for work commensurate with their backgrounds. If required to do what is beyond them they will become discouraged. The teacher should get to know his pupils so well that he could grade them without relying upon any examination, though he may give one for other reasons.

Good teaching also requires catering to the students' psychological needs—giving encouragement to some, putting pressure to work on others and imparting confidence to those who have been defeated by poor teaching in their prior studies. A good teacher must have a knowledge of other psychological problems that beset students, provide advice on how to study, and offer the stimulus of personal contact. To effectively understand his students the teacher must also accommodate himself to the outlook of young people and be part of their world.

A recent study asked about two hundred college sophomores to write 250-word essays describing the *greatest* mathematics teacher they ever had and giving the reasons for their choices. The characteristic chosen by most students, 78 percent, was, "exhibits a genuine personal interest in students." The second highest-ranking characteristic, "conducts interesting classes," was mentioned by only 53 percent. The fifth-ranking characteristic, "knowing the subject," was named by only 34 percent. (This last figure may well be accounted for by the presumption that this attribute is something every student expects of his teacher and so many did not bother to name it.)

Teachers should constantly invite questions from their students. Some researchers are bothered by questions in class that cause them to depart from their prepared lectures or to look up something with which they do not wish to become involved. They prepare "perfect" lectures and regard interruptions as presumptuous or indicative of stupidity.

To present the material properly a good teacher must know how young people think. Can a particular abstract concept be presented successfully to freshmen or should its presentation be delayed until the junior year? Will a rigorous approach to the calculus succeed with students who are beginning the subject? In preparing to teach a course the professor must choose the most suitable text from the dozens of available ones. This selection is time-consuming and calls for judging what students on any one level can handle. Even if a professor spends the time but is not judicious in his choice, he is likely to accept a text that is clear to him but, because of inadequate exposition, unduly burdens the student. He must also pursue and keep track of books and articles on ways of presenting ideas effectively.

Modifications in curricula, usually needed to meet either changing student backgrounds or the introduction of a new area of application, place an additional burden on the teacher. For example, the invention of computers and their constantly increasing use call for altering some courses to take advantage of the availability of these devices. Remedial work with entering college students, now a major problem, requires special courses. The design of new or modified courses is a task that the good teacher undertakes at the expense of considerable time.

A good teacher must be perceptive. He must know what difficulties students have in learning at their stage of the game. It is not sufficient that the professor "knows his stuff." Knowing his stuff too often means that he has forgotten or is oblivious of the obstacles young people encounter in learning elementary ideas. He may regard these as trivial and pass over them as though they did not exist. The alert teacher must also be able to recognize late bloomers and to appreciate that slow thinkers are not necessarily poor thinkers.

Finally, a crucial problem in teaching is motivating the students. It does not occur to many professors that other people with different tastes or perhaps a better sense of values may not like mathematics. Students want to know why mathematics or a particular topic is significant, and not just that their professor likes it.

Even on the more advanced level, where the professor can count on his students having some inclination toward mathematics, presenting theorems without the proper motivation leaves a class with no more than a meaningless collection of theorems, proofs, and procedures. Most professors are interested in mathematics and so cannot appreciate that there is a need to motivate students. In fact, such professors prefer to teach students whose drive is already so well developed that they do not need additional motivation. A professor's greatest joy is the bright doctoral student—and professors love to boast about their bright students as though they made them bright.

Supplying effective motivation is time-consuming and calls for experimentation. It also requires breadth of knowledge that the research-oriented professor does not have and may be unwilling to acquire. The natural and historically valid motivation would be supplied by real, largely physical, problems. But because mathematics professors have abandoned the real world, the break from science has also affected pedagogy.

A good teacher will take a hand in college affairs. Counseling of students and service on faculty committees and the university senate (through which the faculty cooperates with the administration) are obligations of the entire faculty. But researchers do not deign to concern themselves with undergraduate or even graduate affairs that involve cooperation with other departments and the administration. Hence, the burden falls on the men and women who are primarily teachers, and the conscientious teacher does not shirk it.

In short, teaching is a specialty that can be pursued only by people with the willingness to master the art through persistent devotion and experience. It undoubtedly calls for wholehearted and almost full-time effort.

All these considerations, which exhibit the vastly different demands of research and teaching, do not include another vital factor, namely, the personalities of researchers and teachers. A teacher must be able to communicate his knowledge. Do mathematics research professors have this ability? In general they are introspective and introverted; they do not feel at ease with people; they shy away from personal contacts. They like to concentrate on their thoughts. They choose mathematical research partly because mathematics per se does not pose the complex problems that are involved in dealing with human beings. To repeat Bertrand Russell's words, "Remote from human passions, remote even from the pitiful facts of nature, the generations have created an ordered cosmos, where pure thought can dwell as in its natural home and where one, at least, of our nobler impulses can escape from the dreary exile of the actual world." He also said in his autobiography that abstract thinking destroys our humanity and draws us into ourselves, and he confessed that he was drawn into mathematics "because it is not human." Mathematics, then, is a refuge. Research in many fields is a solitary occupation; in mathematics it is likely to be a haven for those with scanty spirit and interests. Would such individuals be likely to communicate successfully with students?

The good teacher must be dynamic, articulate, engaging, clear, warm, sympathetic with students, interested in people as much as in ideas, and even a good actor. He not only should have a lively personality but also must have the proper temperament. He must be interested in young people. Their problems, to some extent, must be his. The student who wants to be an English major and rebels at taking a required mathematics course has a case that to him, at least, seems sound. It is the teacher's function to convince him of the value of that course for his life as an educated person.

There are many professors who, though they wish to be good teachers and approach the work seriously, apparently believe that the wish is the act. The wish may be father to the deed, but the conception, pregnancy, and labor are missing. These require perception, time, and energy devoted specifically to the requirements of good teaching. Much of this art must be learned from the students. To the extent that the students show comprehension, interest, even excitement, satisfaction, drive to learn more, and response to deliberately posed queries intended to stimulate thinking—to that extent the teaching is successful. But where and in what respects it is not must be observed by the teacher and modification be made to remedy defects. This is far more than even the more student-oriented research professors are willing to do, because they reserve most of their time, thought, and energy for their research.

Many researchers are kind, polite, and well-meaning. However, they are also unable to make contact with the students, and some are even afraid to face them. These professors are often quite conscientious. They prepare material carefully; they write it on the board meticulously, and the students spend their class time copying what the professors put there. Such research professors are apparently so unworldly that they have not heard of duplicating machines.

Researchers are usually credited with not only the specific talent that enables them to be creative but also with an underlying superior intelligence that enables them to judge wisely and perform superbly in every situation they may face. The question of whether there is a special talent for mathematics cannot be settled by any objective evidence.

One certainly must doubt whether there are mathematical genes. Insofar as opinion can answer the question of talent, most notable is the opinion of Richard Courant, who was head of the world's leading department of mathematics in pre-Hitler Germany, at Göttingen University, and who converted a nondescript department of mathematics in the United States into one of the country's leading departments, perhaps the leading one. No one knew more great mathematicians of our century, or knew them more intimately. When asked whether there is such a quality as mathematical talent, Courant replied, "There is no such thing." And then he added, "Rather than ask what special qualities mathematicians possess, you should ask, what do they lack that other human beings possess." Perhaps this response also answers the question of whether researchers possess superior intelligence.

Though some administrators might concede a conflict between research and teaching, they fall back on other arguments for insisting on research as the qualification for faculty status. Original research, they say, is the surest proof of intellectual distinction and the surest guarantee that intellectual activity will not cease. Leaving aside for the moment the problem of judging the quality of original research, we must ask what intellectual distinction has to do with teaching. A truly distinguished intellect may be so far above the level of the students he must teach that he will fail to understand their problems and needs, even if he possesses the willingness to do so.

As to original research being the surest guarantee that intellectual activity will not cease, many researchers fade out in their thirties and even more do so in their forties. And if they do, they will most likely be dispirited, depressed, and even sour. Unfortunately, we have no figures on how many researchers fade out nor on how their number might compare with those who emphasize teaching and remain intellectually alive in all sorts of activities. Certainly some research men do burn out. The general belief, though its truth is not definitively established, is that science is a young man's game. As P.A.M. Dirac, a Nobel-prizewinner in theoretical physics, once said, partly in jest:

> Age is, of course, a fever chill
> that every physicist must fear.
> He's better dead than living still
> where once he's past his thirtieth year.

It is vital that professors keep intellectually alive but it need not be through research. There are other ways which we shall discuss later. Moreover, what does it benefit the student if the research professor remains alert but, in view of the nature of modern research, he remains narrow?

Many administrators and chairmen argue that some standard for selection must be used, and since research can be measured whereas teaching cannot, it is better not to risk hiring people who are primarily teachers. The argument is false. We have already pointed out (Chapter 3) that it is almost impossible to secure competent judgment of the quality of research and that in fact most current research is worthless. Evaluation of research boils down to quantity of publication; the contents are irrelevant. As long as his peers accept it, a researcher can publish almost anything—and his peers *do* accept it, because they wish the same treatment. The chairmen and administrators who say that research can be measured whereas teaching cannot are really admitting to counting pages of publication. This measure is as sound as looking at a university's total expenditures to determine whether it is fulfilling its role as an educational institution. One wonders how administrators would have judged the young Isaac Newton, who, after receiving criticism of a paper he had published, vowed never to publish again.

Surprisingly, even in large departments the research cannot be evaluated with assurance. Each person is a specialist in his own field and really knows little, if anything, about the work of his colleagues. Evaluation by committee is no more reliable. Anyone who has worked on committees knows that enlisting several people to share the burden results in each member of the committee expecting the others to do the work and no one contributing anything. Instead, friendship, partisanship, jealousies, favoritism toward one field of research rather than another, gossip, and scraps of information decide the issues. Many department chairmen or committees called upon for one reason or another to judge a faculty member write to specialists in the appropriate field. This recourse to peers is hardly likely to produce a fair judgment. We have already noted (Chapter 3) the defects of the refereeing process, and all of these apply as well to the judgment of published research by peers. Moreover, because fads do play a role, a sound researcher working in a field that has not become fashionable will get a poor rating. Mathematicians are no more discerning of or receptive to new ideas than are small-town politicians whose constituents seem happy with the status quo. The idols of the tribe are not desecrated with impunity.

Beyond judgment by a professor's own department, decisions on promotion and tenure, even if recommended by the department, must usually be approved by a faculty committee. In view of today's specialization of research, there may not be even one member on the committee who can understand the research he must judge. And so the faculty committee, too, falls back on quantity of publication.

How does one judge teaching? Students are possible judges. However, students are young and their mature evaluation may come years later. Moreover, students differ in their needs and demands. The indifferent student may value just entertainment and will rate high a teacher who is only an actor. The mediocre but conscientious student will appreciate steady-going, clear presentations, but flashes of brilliance from the professor will pass him by or upset him. Such a student wants to know only what to do and how to do it. The really bright student, who can read for himself, wants mainly direction, deep insight, and perhaps challenging problems. Obviously, a typical class will contain all types of students and the professor will get a mixed rating. In addition, students are influenced by the grades they get. Finally, if compelled to take a course in a subject they dislike, they are certainly going to be antipathetic to the teacher. Hence, evaluation by students, though it should not be ignored, cannot be the main basis for deciding on a teacher's effectiveness.

Admittedly, teaching is difficult to evaluate, and good teaching takes many forms. The professor who stimulates and excites students to work for themselves—even when he does little to present the subject matter proper and is even unsystematic in doing so—is certainly a good teacher. The sober, careful expositor who helps students to master the material and gives them the confidence that they will do well, but presumes student drive, is another excellent type. Still another is the professor who may be a dud in the classroom but who builds close personal relationships with his students, spends hours with them on their individual needs, and gives each student the feeling that he has a friend and counselor.

A department chairman can do a great deal to evaluate a teacher. He can make it his business to have frequent discussions on what topics the professor chooses to teach in a given course and how he presents his course. The chairman can get to know a teacher's character and temperament, his interest in teaching, his choice of texts, the kind of examinations he gives, and his availability to students. Teachers can themselves demonstrate expertise by writing a good text, an expository paper, or developing a better pedagogical approach to a topic.

However, there is no objective and certainly no quantitative measure of teaching ability. Judgment enters here as in the evaluation of research and as, in fact, in all the important decisions men are called upon to make. And administrators who are not qualified to exercise judgment should not be allowed to serve in positions where they will be required to do so. Around the turn of the century William James said that the social value of a college education was to be able to recognize a good man when you saw one. But this ability is not possessed by most administrators. To decide whether a teacher is good they ask him to supply letters of recommendation (usually written by those who are no better judges than the administrator), and they ask for a list of his publications. Despite the inescapable need for applying true judgment instead of resorting desperately and fruitlessly to quantitative data, many administrators are unwilling or fearful to apply it. Judgments may be wrong, but all one can expect of a good administrator is that he is right most of the time.

The saddest fact about the evaluation of teaching is that most administrators don't care whether a teacher is recognized as superb by colleagues and students. Teaching just does not count in the universities. Of course administrators deny this.

All things considered, the very nature and demands of research and teaching are so diverse, the personalities and abilities required so different, that only a rare individual with an abundance of energy can undertake to excel in both areas—at least during the same period of his life. Generally, good performance in teaching is in inverse proportion to efforts in research. Of course some research men may be willing to teach, but ability to do so is another matter. Their wisdom may be consumed in their research.

Though any attempt to compile statistics on the number of mathematicians who were or are both good researchers and good teachers would be a massive task with much room for misjudgment, it is significant that two of the three men commonly selected as the greatest mathematicians present relevant case histories. Newton was a professor at Cambridge University and was known to be a very poor teacher. At times no students at all attended his lectures, and of those who did attend, few understood him. The paucity of students did not disturb him, but he was almost paranoid in his concern to receive credit for his creations.

Though not a bad lecturer, Gauss did not care to teach and said so. In 1802 he wrote to the physician and amateur astronomer Wilhelm Olbers:

> I have a real aversion to teaching. For a professor of mathematics it consists of eternal work to just teach the ABC's of his science; of the few students who go on, most continue to gather a file of information and become only half-educated, whereas the rare gifted students will not allow themselves to be educated through lectures but instead learn by themselves. And through this thankless work the professor loses his precious time.

Gauss attracted only a few students to his lectures, whereas his colleague, Bernhard Friedrich Thibaut, who contributed very little to mathematics, had a hundred. Gauss gave the same lectures with very little variation, year after year.

The third of the "greats," Archimedes, was not a professor

and, moreover, lived at a time when education was conducted so differently that his case is irrelevant.

In recent years not many great researchers have expressed in print their attitude toward teaching. But Godfrey H. Hardy, one of Great Britain's leading professors in the first half of this century, did so in his *A Mathematician's Apology*. "I hate 'teaching,'" he wrote, "and have had to do very little, such teaching as I have done having been almost entirely supervision of research; I love 'lecturing' and have lectured a great deal to extremely able classes; and I have always had plenty of leisure for the researches which have been the one great permanent happiness of my life." One understands why Hardy titled his book *A Mathematician's Apology*.

In our own time the researcher's lack of interest in teaching is patent. When a university seeks to attract such a person, it offers not only money, but also a light teaching load—the lighter the better. The university officials know that researchers do not want to teach. Beyond the small number of teaching hours, the researcher is offered the privilege of devoting some of those hours to a course or seminar in his own specialty. Many researchers have said frankly, "Universities would be fine places if there were no students." Quite a few have expressed their contempt for undergraduate teaching and often can be heard at evening social gatherings to bemoan the fact that they must meet an undergraduate class the next morning. They sneer at the mere teacher and act condescendingly toward the students whom they regard as dull and unworthy of their attention. Such researchers may know the mathematical theory of ideals, but they are certainly not familiar with the ideals of teaching. They present familiar techniques mechanically, and when faced with the inevitable consequence that the students are bored and baffled, they criticize the students for lack of interest. Actually, it is the professors who are not interested. By continuing to teach the same old techniques, they avoid thinking about how to present the material. Admittedly, there are also people engaged primarily in teaching who would be graded A only for the speed with which they can empty a classroom. A bad teacher is a small-scale disaster that wreaks havoc as long as he lives.

Some would claim that ego and approval of the peer group are factors favoring the pre-eminence of research. Darwin did say, "My love of natural science...has been much aided by the ambition to be esteemed by my fellow naturalists." Yes, scientists do love recognition by their peers, especially since the time of Pythagoras. For mathematicians the peers consist, reasonably enough of fellow mathematicians—a society conditioned by the universities to approve only research. If the peer group were encouraged to grant recognition to teaching it might in fact do so. In any case, ego and approval from the peer group need not impel a frenzied rush to publication. True ego would demand genuine accomplishment, even if the attainment of it were to take ten years.

In view of the conflicting demands of research and teaching, how can responsible men continue to affirm that research professors are automatically good teachers and that good teachers ought necessarily to be seriously involved in research. It is understandable that researchers cannot believe that they are not good teachers. It would hurt their pride; and they are proud to the extent of egotism. Moreover, since so many are naive about what lies outside their own research, they uncritically accept these tenets. They would be shocked if they were to request their students, at the end of a course, to answer a questionnaire asking them to check one of the following: The teaching in this course has been a) excellent, b) good, c) fair, d) poor, e) execrable. But their confidence in their teaching ability is so firm that they do not see the need to question their own effectiveness.

One would think that chairmen of departments would be more responsive to their obligations to the students, but they are not. Since chairmen are anxious to show that they can build and maintain research strength, they favor researchers. Thus, the chairmen, too, sacrifice the students. Many chairmen prefer researchers because they know only research and with unconscious immodesty seek men in their own image. Teaching they do not understand and so they regard it as unimportant.

It is more surprising that administrators accept such crass doctrines without evidence. It is their responsibility to foster teaching and research, and they should know whether researchers can also supply the teaching needs of students. But the administrators of even a moderate-sized university are far removed from the activities of the individual departments, and so they are unable to judge the quality of the faculty, the reasonableness of the extent or quality of the offering, and the quality of the teaching, even if that should be a concern. They concentrate on budgets and maintenance of regulations. A good bookkeeper could do what many deans and vice-presidents do. University administrations are staffed largely with people who strive hard to perpetuate what they do not understand. In such matters as teaching and research they are naive and merely repeat what they have heard. These administrators, too, do not see the need to check their beliefs.

Many administrators who know that research is in direct conflict with teaching profess nevertheless that research is a prerequisite. Why? What these administrators really seek is prestige, and any measure that builds prestige is favored. Whereas fifty and more years ago they sought to obtain it by attracting socially elite students, in today's world the medium is research. Since teaching can be appreciated only by students, whose opinions do not count in the adult world, teaching is not valued. Researchers, on the other hand, publish, and so their names and their university affiliations receive publicity. Because research is considered (certainly by those doing research) to be a mark of genius, it is accorded glory that reflects on the universities sponsoring it. It is the mink coat of the university world. Researchers win Nobel prizes, awards from professional societies, election to the National Academy of Science, and other such honors, and these, too, build up the prestige of the universities employing them. In *The American University* Professor Jacques Barzun mockingly suggests that universities have used the formula

$$Vip = \frac{2p + 5n}{f}$$

or Visible Prestige equals twice the number of Pulitzer prizes plus five times the number of Nobel prizes divided by the total number of faculty to measure their prestige. Of course this formula will not quite do for mathematicians; their work is not awarded Pulitzer or Nobel prizes.* However, election to the National Academy of Sciences or high office in a professional society may substitute for Nobels and Pulitzers in Barzun's formula. It matters not whether the honors are deserved (see Chapter 11); if the university can parade them in its literature, or if they receive five-line notices in the newspapers, that suffices. It apparently has not

* In an article in the *New York Times Magazine* section of November 21, 1976, Saul Bellow, winner of the Nobel Prize for Literature in 1976, quotes his wife, Alexandra, a professor at Northwestern University, to the effect that Nobel excluded mathematicians because his wife's lover was a mathematician.

yet occurred to administrators that a more effective means of building prestige would be to hire a Madison Avenue advertising firm, but no doubt that soon will be done.

In their unbridled quest for prestige the universities do not rely solely upon researchers. The professor who heads a large project, such as space exploration, who travels frequently to Washington, or who receives newspaper attention for whatever he does is as desirable for faculty status as the most competent research man. So also are high government officials who can still in some way be named as faculty though they do nothing to further the legitimate activities of the universities.

There is ample evidence that universities seek prestige rather than quality. University Y has a fine mathematics department but nevertheless offers Professor A of University H a handsome salary to leave H and join Y. Now whatever good work A may be doing he can do just as well at H as at Y; hence the world does not benefit by the change in locale. But if A has prestige that will accrue to Y, that suffices for Y to invite him. Universities give all sorts of reasons for attracting people from other institutions. Of course, if A's subject is basic and there has been no one to teach it, or if a college decides to add graduate work and wishes to build a research faculty, the offer is justified. But these sensible reasons are much less frequently operative than the desire to acquire or enhance prestige. Universities hire professors the way some men choose wives—they want ones that others will admire.

The drive for prestige is evident almost everywhere. One university president making his maiden speech in that capacity to a group of faculty, alumni, and friends of the university opened his speech by mentioning the Nobel-prizewinners, National Book Award winners, and other prominent figures who were members of the university staff. This opening was evidently intended to impress the audience. At no time during the speech did the president mention any notable feature of the teaching function of the university. Ironically, this same university had reduced its retirement age to sixty-five and therefore, in effect, dismissed some very capable men and women, while continuing to appoint prestigious outsiders who were past sixty-five.

Why do the universities seek prestige? The main reason is money. Money has become the overriding concern of administrators. We have already noted that researchers attract government and private foundation money—which is then spent to attract more expensive researchers. Private universities also compete for students because tuition covers a substantial part of the universities' expenses, and students, often on their parents' urging, like to attend prestigious institutions. State universities compete for students so that they can demand more money from their legislatures. But there are intangibles also. Universities are run by individuals, and the prestige of the university enhances the individuals' standing. As Thorstein Veblen remarked many years ago, "Men of 'affairs' have taken over the pursuit of knowledge." Even professors, supposedly wholly devoted to their work, will accept a lower salary at a more prestigious university just to be more visible under the brighter radiance.

Though the craving for prestige is the motivating force behind most university actions, research is the socially approved word and, in fact, is one effective means of acquiring and maintaining prestige. Hence, the university administrators in their speeches and writings use the term "research" even when they are fully aware that it is a euphemism for what they really mean.

The universities' drive for prestige puts the professors in the anomalous position of being hired to teach, required to do research, and prized for the prestige they accord to the university. The professors react accordingly. Even those who prove to be mediocre in research persist in trying to base their claim for advancement on research, partly because they know that this will count more than any other activity and partly because they are not convinced that their research is minor. They still hope to write the papers that will show their greatness and so are unwilling to devote appreciable effort to teaching and to university and departmental concerns. Others seek visibility in circles that university administrations prize, no matter what is sacrificed in the process.

The present qualifications for getting ahead in the professorial career are neatly summarized in the statistics of responses made in 1964 by members of the American Political Science Association:*

ATTRIBUTE	RANK
Volume of publication	1
School at which doctorate was taken	2
Having the right connections	3
Ability to get research support	4
Quality of publication	5
Textbook authorship	6
Luck or chance	7
School of first full-time appointment	8
Self-promotion ("brass")	9
Teaching ability	10

The inordinate emphasis on research is a relatively new phenomenon in American universities. We have related (Chapter 2) the fear expressed by Charles W. Eliot in 1869 that research would detract from teaching. Though Eliot may have exaggerated the danger in the late nineteenth century, inasmuch as research in the United States was then

* Somit, Albert and Joseph Tannenbaum: *American Political Science*, Atherton Press, 1964.

in its infancy—and though he did later recognize that the universities must also support research—his fear ultimately proved to be justified.

Since the late 1940s the universities have demanded research and only research as the sole criterion for acceptability in the professorial career. This preference reminds one of G.K. Chesterton's remark, "My mother, drunk or sober." The mania for research has produced an invidious system of academic promotion, perversion of undergraduate education, and contempt for and flight from teaching. As Marshall McLuhan would say, the medium is truly the message, and the medium for university mathematics education is totally unsuited for the message. Or, as another professor put it at a recent meeting of the American Sociological Association, "Teaching represents the big vacuum in higher education."

5

The Debasement of
Undergraduate Teaching

But they that spread a feast all for themselves eat sin and drink of sin.

The Bhagavad-Gita

Research is expensive. Research professors command high salaries but cannot be asked to teach many hours. And they teach mainly graduate courses and seminars in which the number of students is small so that the tuition does little to defray the expenses incurred. In disciplines that require laboratories the cost is still greater. Much of the research conducted by professors does not involve students at all and so must be supported entirely by the universities, though governmental and foundation grants do cover some of the cost.

In contrast to research, university undergraduate education is profitable. Students in the private universities pay a high tuition. In the public universities state and city funds plus tuition provide the equivalent. The student body is large; hence the income from undergraduate students is considerable. Why not, then, divert money obtained from undergraduates to support graduate education and research? This is precisely what universities do.

To "handle" undergraduates cheaply—it would be factually incorrect to say "teach" them—the universities have adopted two measures: large lecture classes and the use of graduate students as teachers. Lecture classes, usually taught by research professors, may contain as many as a thousand students. In a few universities some of the students hear the lecture only through closed circuit television, with two thousand to five thousand students hearing one professor.

The objections to large lecture classes, in mathematics at least, are almost evident. Teaching is not solely a matter of delivering material. Students differ in background and ability, and effective teaching calls for knowing the difficulties of each student and being able to resolve the questions each can pose. Questions and answers are more important than a professor's repetition of what can be found in books or what can be put in lecture notes, duplicated, and passed out to the students. Moreover, mathematicians particularly claim to teach thinking. This calls for eliciting responses to stimulating questions, encouraging intelligent suggestions from the students, discouraging poor suggestions (though not to the extent of destroying confidence), and guiding the students to a correct proof or answer by continual probing of their thoughts. Clearly, this cannot be done in a large lecture class.

One defense administrators proffer for the use of large lecture classes—that it permits many students to listen to a famous professor—amounts to just what it says. They can listen and perhaps be kept awake by a booming voice or good acting, but not necessarily learn. At best, a good lecture in mathematics is systematic, complete, correct—and dull. More often it is a monologue, if not a soliloquy; teaching calls for dialogue. The worthlessness of the lecture system is pointedly described by Alfred North Whitehead in *The Aims of Education*: "So far as the mere imparting of information is concerned, no university has had any justification for existence since the popularization of printing in the fifteenth century." The lecture system is the instrument of last resort and actually an evasion of teaching.

To supplement the large lecture sessions, which some universities recognize as inadequate, students meet one or two hours a week in small groups with a graduate student. The graduate student answers questions, goes over homework, and usually gives the quizzes and grades. Were the graduate students really equipped to do the tasks that the lectures fail to do, the lectures could be dispensed with. But guiding a student to think for himself is an art that can be learned only after years of experience. To recognize and advise on psychological problems as well calls for maturity the graduate student does not have. He also cannot afford the time to keep up with educational trends, and so does not know what background he can or cannot expect of his students. Moreover, he is, naturally, so much concerned with his own progress toward the doctorate that he will almost surely devote as little time as possible to his pedagogical duties.

Where large lecture courses are supplemented by small discussion or problem sections, the professors heading such courses give little or no attention to the graduate assistants who conduct these small groups. Having delivered their lectures, the professors believe that their responsibility for the course has ended. Faced with such indifference on the part of the professors, what inspiration and incentive do the assistants acquire to take their own teaching duties seriously?

Some universities rebut all charges of poor teaching, in particular the charge that they use large lecture classes, by citing the high faculty-to-student ratio. This is a meaningless figure. Many prestigious professors who "grace" the campus teach either not at all or very little. A more meaningful number would be the ratio of faculty teaching hours to student course hours, and even this ratio would not describe the quality of the teaching.

A major two-year study at a very prestigious university to determine what improvements in undergraduate housing could be made uncovered that students did not seek better food, more spacious rooms, or even more members of the opposite sex under the same roof. What they did want most was more personal contact with their professors.

Perhaps the value of the lecture system is evidenced by one wag's story whose point should not be ignored. A professor at a major university was called to Washington so often that he ran into conflict with his lecture schedule. He decided that he would tape his next lecture in advance and have the tape run off for the class. He happened to return to the university earlier than he had expected. Passing his lecture room, he heard his taped lecture being played to the class. Curious to see how the class was reacting, he entered the room. To his surprise there were no students in the room—but on each seat there was a tape recorder. Lectures delivered by tape are, of course, uncommon. But insofar as effectiveness is concerned, they are no different from lectures delivered by professors in person. Large lecture classes and the computerized handling of student registration and records have dehumanized education. It is no wonder that the Berkeley students complained during the period of strong protest in the late 1960s, "I am a human being; do not fold, spindle or mutilate."

The second maleficent cost-cutting practice of the universities is to assign students to small classes of thirty or so with a graduate student, known as a teaching assistant, or intellectual dishwasher, doing all the relevant work—lecturing, grading, advising, determining course content and other matters. These novices are used even to teach the special courses for prospective elementary and high school teachers.

In some universities as many as one hundred to two hundred graduate students do undergraduate mathematics teaching. In 1972 there were 109,000 teaching assistants in all departments of about 250 universities. It is estimated that about 50 percent of the freshman and sophomore mathematics courses in large universities are taught by graduate assistants. Ironically, the very same universities that insist that good teachers must be researchers use graduate students on a large scale.

The reliance upon graduate students as teachers is underscored by events of the last few years. The number of graduate students is declining sharply because jobs in the academic world are now hard to get and fewer students are training for the doctorate. This does not worry the graduate schools so much as the fact that the number of graduate students serving as teaching assistants will not be sufficient to cover the undergraduate classes. Far be it from the minds of administrators to ask professors to teach the elementary undergraduate courses, even though beginning college students need mature teachers more desperately than do advanced undergraduates. Instead, chairmen of university departments canvas the undergraduate schools for bright seniors who might enroll as graduate students and thereby supply an adequate number of teachers.

Many large universities use not only graduate students but also adjuncts (housewives returning to teaching, high school teachers teaching at night, and full-time teachers at other institutions teaching for extra pay). Clearly, these people cannot really immerse themselves in the work. Yet two-thirds of the total number of undergraduate mathematics courses are taught by graduate students and adjuncts.

The graduate student, naturally concerned with his career, will devote himself primarily to his own studies and give short shrift to his teaching. He will hurry through poorly prepared lectures, discourage students from seeing him in office hours, make up poor quizzes and final examinations, and grade almost arbitrarily. Covering the syllabus is a must and the "poor" students who can't maintain the pace are left to fend for themselves. After all, they must be lazy or stupid. They, the graduate students who "made" it, didn't have any such failings and presumably no other cause, such as a student working 20 or 30 hours a week to pay his tuition, enters their minds. If they are permitted to pick texts, young people, hardly aware of the true goals of education or even of mathematics education, pick abominations. Should the teaching assistant feel conscience-stricken, he can ease his conscience by noting how his professors handle their graduate teaching. If, on the other hand, he does do all that conscientious teaching requires, he may never complete his degree—or he may complete it with a poor thesis and not get a job in any reputable institution.

Teachers must have wisdom and wisdom is acquired from experience, not books. It is not a criticism of graduate students that they have not acquired wisdom. It is a criticism of university administrations that they depend upon the widespread use of graduate students and, in most universities, make no formal attempt to teach these young men and women the art and skills of teaching, whether they be the presentation of material, composition of examinations, grading, or advising young students. If a graduate student should, on his own initiative, take a course in pedagogy, he would undoubtedly lose the respect of his mathematics professors.

A few of the universities that use graduate students as instructors do try to train them to teach. But such efforts are as likely to be successful as efforts to make water run uphill. Since most universities do not appoint or retain professors who are good teachers, who will train the graduate students? Moreover, as we have already noted, graduate students know that their futures depend primarily on securing a Ph.D. Hence, they must give this activity priority. Their success or lack of success as teachers will not count. Finally, what incentive would a graduate student have to learn how to teach when he knows that the very professor who may be advising him is almost sure to be one who is indifferent to teaching?

The use of graduate students to teach freshman and sophomore courses is especially reprehensible. Beginning students find the transition from high school to college difficult, even traumatic. Many have not learned how to study, how to use a library, and how to live in a strange environment. Freedom from parental guidance leaves some desolate and others irresponsible. During the first year especially students need help and advice, and it is during that year that they are put into the hands of a graduate student who has yet to grow up himself.

Allowing graduate students to take complete responsibility for conducting courses is sometimes defended on the ground that there are not enough qualified professors to handle the mass of students. This is not true; good teachers for undergraduate courses have been in plentiful supply for the past twenty or thirty years.

The use of graduate students is defended on other grounds. They can earn money to support themselves while studying. Many do need to earn money, but it should not be at the expense of the undergraduates, who are paying for and need high-grade instruction. Another defense is that graduate students, many of whom will become professors, need to acquire experience in teaching—to learn by doing. One wonders whether administrators who assert this would go to a first-year medical student to cure an illness. Graduate students should acquire experience, but they can do so in ways that do not sacrifice the undergraduate. Graduate students can observe and can serve as informal tutors to help students having difficulties. They can give a few lectures in the presence of the professor, who can then offer constructive criticism.

Still another argument for the use of graduate students as teachers is that they are more enthusiastic about teaching than many of the professors. But if their enthusiasm outweighs their shortcomings so that they can serve as well as professors, then the correct interpretation of that argument is that many professors should not be professors at all.

If their alleged enthusiasm does not suffice to justify the use of graduate students as teachers then surely as young people they would have empathy with the undergraduates. Quite the contrary. The graduate student is competent in mathematics, else he would not undertake graduate study. On the other hand, since 95 percent of the undergraduates are not very successful in mathematics, the graduate student is most likely to be contemptuous of those who have difficulties in mastering what he took in stride.

The use of graduate students as teachers has been challenged in the past, and one defense offered by administrators is that tests show no difference in results whether undergraduates are taught by graduate students or by professors in large lecture classes. Of course, tests do not measure inspiration, insight, pleasure or displeasure, the perhaps unnecessary hardships students have to undergo to achieve performance, and the psychological lift or damage to the student. But let us for present purposes accept tests as a measure of teaching effectiveness. The results cited by administrators prove nothing. Large lecture classes are as bad a teaching measure as the use of graduate students; hence the test grades are equally bad. A significant difference might be expected between students taught in small classes by professors and those taught by graduate students. And if, in fact, the small classes were taught by professors who are *teachers*, the tests would show which type is superior. However, for the test results to be significant a large number of students should be involved. But few, if any, universities have mature teachers to handle a sizable number of small sections. Given poor teachers, even students taught in small classes are not likely to do better than those taught by graduate students.

Entrenched professors, who benefit in lighter teaching loads and high salaries, also defend the use of graduate students and proffer the same self-serving arguments as do the administrators.

Very belatedly, in 1975, the American Mathematical Society and the Mathematical Association of America set up a joint committee on the "Training of Graduate Students to Teach." In view of their immaturity, their relative ignorance of content and pedagogy, and the pressure on graduate students to earn a Ph.D., training graduate students to teach will be as effective as training a six-year-old to run a four-minute mile.

Are the universities being fair to undergraduates? Tuition in private universities, and tuition plus state and city support in public universities, do more than meet the costs. What the universities prefer to do is to save money on the undergraduate instruction and spend it on high salaries to attract high-grade or prestigious research professors. Research is also a legitimate function of the university, but it is carried on by cheating the undergraduates of their due. To use large lecture classes conducted by professors who are unwilling or unable to teach, or to use inexperienced, immature, and preoccupied graduate students is unethical and amounts to robbing Peter to pay Paul.

How do the universities get away with such practices? They use the prestige acquired by having big names attached to their staffs to attract students; and since American parents want their children to benefit professionally and socially by having them graduate from prestigious universities, they continue to send their children to these places. The public universities offer the added advantage of low tuition, and though they are the worst offenders in the use of graduate students, poorer families have no choice but to send their children there.

In contrast to the universities, the four-year colleges of this country do a far better educational job. On the whole the teaching is conducted by mature professionals, and the classes are small. Unfortunately, this assertion must be qualified. Many of the four-year colleges were founded one hundred to one hundred-fifty years ago by church-affiliated groups so that parents could send their children to schools that would reinforce rather than raise doubts about the truth of their particular religious beliefs. To be sure, the religious ties of the denominational colleges have now become nominal and in many cases have been dropped altogether. But so has financial support from the churches. Also, when travel was more difficult it was undoubtedly necessary to found many small colleges so that they would be accessible to students. These colleges with enrollments of five hundred to one thousand students cannot afford to compete for faculty, to offer an adequate variety of courses, and to maintain adequate facilities, notably a good library. They cannot survive, and the only recourse is consolidation with nearby institutions. Many have gone out of existence.

Nor is life easy for the larger four-year colleges. Endowments and state support that were adequate twenty and fifty years ago no longer suffice. When jobs are scarce the four-year colleges can secure well-trained faculty despite higher teaching loads. But many of these recruits have their eyes on university positions where more stimulating contact with colleagues and better library facilities are available. Therefore, although not pressed to do research at the colleges, they will nevertheless devote most of their time and energy to research in the hope that their work will ultimately attract attention and that a university appointment will be forthcoming. The lure of the lucrative research position is almost irresistible.

Moreover, qualified teachers who accept appointments are generally subject-oriented. Their training and interest are in the subject they undertake to teach, and they continue to seek approval from others in their own discipline. And having been educated primarily by professors who lecture and who presuppose an interest in the subject, these undergraduate teachers tend to do the same thing. Nor does it occur to the usual undergraduate professor, however intelligent, that he must know what background high school students bring to college. Indeed, many of the concerns that should be foremost in the minds of good college teachers are too often recognized either belatedly or never. This is understandable. Not only did they receive no formal instruction in pedagogy, but rarely is there a serious discussion among members of a college department about the problems and techniques of pedagogy; and only extremely rarely does an administrative official undertake to institute such a practice.

Quite often it is not the professors at the four-year colleges who are responsible for poor education. Many of the administrators are not content to play a vital, if less prestigious, role in education. They seek to emulate the universities and demand research as a condition for employment and advancement. Nevertheless, despite all these shortcomings, the better four-year colleges are still the only institutions in which the intent to teach is uppermost. Fortunately, many competent mathematicians who prefer teaching and the congeniality of friendly colleagues to the fierce competitiveness of the large universities are willing to sacrifice prestige for a useful and satisfying contribution to society; they do gladly accept appointments to four-year colleges and devote themselves to teaching. Thanks to them the four-year colleges are the mainstay of undergraduate education. Unfortunately, the universities train twice as many bachelor candidates as the colleges.

Less satisfactory than the four-year colleges are the two-year junior and community colleges. Many are now taking advantage of the scarcity of jobs to get Ph.D.'s from prestigious institutions. In fact, some are insisting that all new faculty appointees have a Ph.D. But the teaching problems of the junior colleges are even more severe than those of the four-year colleges, because all high school graduates are automatically admitted. Ph.D.'s, trained to do research, are not suitable faculty for the two-year colleges.

The low quality of the teaching in most of the institutions wreaks havoc in many directions. The research work done by most professors, as we have already noted, is overvalued. A teacher, on the other hand, influences during his career thousands of young people, many of whom will be leaders in various areas of national life. Their attitude toward education, their decisions about what careers to follow, and their confidence in themselves are determined, or at least strongly influenced, by their teachers. Many a young person has abandoned a scientific career because mathematics proved to be an obstacle, and one may be quite certain that in most of these cases the teaching was the true source of the trouble. Let us recall also that the universities train the high school and elementary school teachers. Hence, what the universities do affects teaching at all levels. (See Chapters 8 and 9.)

Though students are the chief victims of poor mathematics education, the fate of mathematics itself is also imperiled. The mathematical strength of any country or civilization depends not only on the geniuses who produce brand-new worthwhile results. Indeed if these geniuses were isolated and unappreciated their output would be voices in the wilderness. Mathematical knowledge must be kept alive and transmitted to younger people. Even a scientific breakthrough is of questionable value in a society that does not know how to absorb it. To recognize and properly educate even the future mathematicians, the educative process must be widespread and many good teachers must be available. Talent is found in the most unexpected places, and unless a teacher is competent and alert it will be overlooked again and again. It was an obscure teacher who recognized the ability of Gauss, one of the world's greatest mathematicians, and who recommended that the Duke of Brunswick support him. Gauss's father, a bricklayer, certainly could not have sent him to a university. Whether or not good teachers themselves contribute to research, they are sowers of seeds that take root in the minds of young people, some of whom will be Gausses and others of whom will enable society to profit from the work of the Gausses.

Despite the importance of teaching, an importance far greater than most research, the good teacher has no place in the present-day university world or even in those four-year colleges that insist on research as a requisite for faculty status. Thus, in higher education the teacher, however excellent, who refuses to bow before the gilded idol of publication, finds fewer and fewer outlets for his talent. In an atmosphere where research alone is valued, the teacher is made to feel that he is a failure, that he really does not belong but is tolerated as a second-class citizen because by some oversight he acquired tenure. For him promotions are slow and salary, where not legally prescribed, remains far below that of the research man. He is regarded as a drag on the department, which somehow has been stuck with him. As one writer about the current scene put it, college professors are now divided into two classes, entreprenurial professors and teaching professors—or winners and losers.

In the most prestigious universities, and in many of the would-be prestigious ones, there is not a single tenured mathematician who is there primarily because he is a teacher. Administrators sometimes admit this but claim that instead they offer an intellectual climate and an opportunity for undergraduate students to mingle with intellectually challenging fellow students. But all of this is an indigestible ersatz for the food that students should be receiving from competent teachers. Moreover, if the teaching is really unimportant, why should tuition pay for faculty salaries? Ironically these same administrators want their own children to have good teachers and don't hesitate to speak up when their children fail to get them.

In contrast to teaching, research has acquired a sanctity and research professors an air of sanctimoniousness even when their papers are rubbish, though not clearly recognized as such through their veneer of terminology and symbolism. To do research is nobler and a higher mission than to teach. Communicating knowledge to one hundred or one hundred-fifty students a year is not valuable. But a paper in a journal which no one may look at is. Significant communication does not rank with insignificant scribbling.

Tenured professors with reasonably good salaries do have a measure of freedom. They need not publish a large number of pages per year and can devote part of their time to students without risking much. However, professors are a cross-section of human beings. The thousand-dollar raise that might be obtained if one does more research does beckon. The accolades of colleagues and professional organizations are sweeter than the accolades of students, and the prestige of research outweighs the prestige of great teaching.

Curriculum and requirements for the bachelor's degree are being reexamined in many universities and colleges today. Deans and professors who tried to appease students during the protests of the 1960s by abolishing almost all requirements of specific courses for the degree are now convinced that the liberalization went too far. Of course, curriculum and degree requirements are important.

But these reforms are not the most important ones that should be undertaken. The best curriculum, whether for required or optional courses, will be manhandled under present teaching conditions. The research professors will continue to ignore bulletin descriptions and teach what they please, and the graduate students will bungle thoughtfully planned subject matter because they are not qualified to teach.

The same remarks apply to the continual discussions about lines of command, delegation of responsibilities, review procedures, and all other organizational problems of college and universities. An automobile manufacturing company may have an efficient organization and blueprints for the ideal car, but if the available gasoline is poor, the organization and blueprints are all but worthless.

6

The Illiberal Mathematician

Education is not in reality what some people proclaim it to be in their statements.

Plato

The liberal arts colleges—whether integral parts of universities or independent entities—now cater to a variety of student interests. Adequate instruction calls for courses that, while conforming to the values and objectives of a liberal arts education, serve these interests. Mastering the requisite materials and incorporating them in suitable pedagogical format constitute the major tasks of educators.

But the professors are under pressure to do research, and the graduate students, to obtain a Ph.D. Both are narrowly educated, though in different respects and for different reasons. Thus, both groups of teachers, whether engaged in or training to do research, know some mathematics but are ignorant in science and pedagogy, and both are concerned almost entirely with content and values that only mathematicians prize. Yet both must tackle the courses that are supposed to serve many different student interests. Just

112   Morris Kline

what is it that professors and graduate students, taking their cue from professors, teach in some of the most fundamental courses?

The largest single group taking mathematics in the colleges and universities consists of liberal arts students who register for a mathematics course only to meet the requirements for a degree. The typical registrant is indifferent to mathematics or actively dislikes it. Nevertheless, the teaching of these nonspecialists is probably the most important task of the professional mathematician. It is important from the broad sociological point of view and even from the standpoint of mathematics, because many of these students will become leaders in our society and will decide how much support to give to the subject. Also, since the segregation of students into prospective users and nonusers of mathematics is based on interest rather than ability, the nonuser group contains some of the most worthwhile students.

To cater to the nonusers most colleges, whether independent or part of a university, offer what is called a liberal arts mathematics course. Despite the importance of this course, most mathematicians despise teaching it. It is beneath their dignity to bother with nonmathematicians and to waste their precious time and knowledge on mathematical "nonentities." However, many professors are obliged to teach the course in order to fulfill the required number of teaching hours. What, then, do they teach? Because most mathematicians are narcissists, they offer reflections of themselves. They prefer pure mathematics, and even in that area only topics that strike their fancy. Let us examine some of these topics.

Their favorite topic by far is the theory of sets. A set is no more, of course, than a collection of objects. But students are asked to learn specific operations with sets such as union, intersection, and complementation, and are also asked to learn properties of these operations. Set theory also includes infinite sets, such as the set of all even integers. All this material has no, or at best trivial, bearing on elementary mathematics or on real phenomena. In addition, the concept of an infinite set baffled and was rejected by the best mathematicians until the 1870s and is still unacceptable to many today. Since the study of sets, and particularly infinite sets, does not return even infinitesimal riches to the student, he does not see why he should attempt to make his reach exceed his grasp.

Another favorite topic is the theory of numbers, wherein one studies unusual properties of the integers. Some numbers, such as 6, are "perfect" because each is the sum of its divisors (other than the number itself), as 6 is the sum of 1, 2, and 3. "Unfortunately" there are very few known perfect numbers; so the course soon proceeds to study prime numbers. A number is prime if it is divisible only by itself and 1. Thus 7 is a prime, whereas 6 is not. Any number of theorems treat properties of prime and nonprime (composite) numbers. To some mathematicians prime numbers are delightful, intriguing members of the number system. To the students they are hostile strangers. When they learn that there is an infinity of prime numbers they become convinced that the world is full of enemies.

The theory of numbers also includes what is called the theory of congruences, and this topic seems to be a must in liberal arts courses. The theory of congruences concerns an arithmetic that is suggested by how our clocks record time. Six hours after nine o'clock the clock reads three o'clock; that is, $9 + 6 = 3$. In other words, twelve and any multiples of twelve are discarded or counted as zero. Certainly this is a peculiar arithmetic. The theory of congruences studies many such varieties. But even ordinary arithmetic is dull, however practical the knowledge. What, then, can students find interesting in clock arithmetic? Moreover, since many are still insecure about the operations of ordinary arithmetic, the new arithmetic shatters what little confidence they had.

A favorite topic in liberal arts courses is axiomatics (see Chapter 3). Because the axiomatic approach to teaching is now the popular one in many courses, this topic warrants special attention here. Every branch of mathematics is founded on axioms, the prototype being Euclidean geometry. There is much to be gained from a study of the axiomatic basis of any branch. Indeed, the most momentous development of the nineteenth century—non-Euclidean geometry—resulted from a change in the axioms of Euclidean geometry. But what features of the study of axiomatics are presented in a liberal arts course? Some professors introduce a system of axioms and merely discuss what properties the system should possess. One of these, for example, is independence; that is, it should not be possible to prove any one of the axioms on the basis of the others, for in that case the provable axiom is more properly a theorem. The study of properties of an axiom system is of great interest to specialists in the foundations of mathematics, but to tyros the axioms are the least significant part of any mathematical development. They are the seeds from which the fruit eventually emerges. Hence, a treatment of the properties of the axioms themselves has little value.

Another type of "play" with axioms is to show students that the set required to develop a branch of mathematics is not unique. One can change the axioms and still deduce the same body of theorems. Or one can in some branches reduce the number of axioms, though this may involve the need for more complicated proofs of the theorems. The latter activity is relatively simpleminded and usually profitless. One is not surprised to find, then, that students are not exhilarated when they are shown than ten axioms can be replaced by nine or nine and one-half.

One truly liberal arts value to be derived from the study of axiomatics is to help people become conscious of their actual assumptions when they make a decision or adhere to a belief in any sphere. But this "carry-over" of mathematics education to other areas of our culture is never mentioned. The teaching of axiomatics per se reinforces the contention that mathematics does *not* teach critical thinking. For if it did, professors would certainly ask themselves why they teach axiomatics in a liberal arts course.

Some liberal arts courses and certainly the more advanced courses do indeed derive the theorems that are implied by the axioms, just as the high school course in Euclidean geometry does. This deductive approach to a branch of mathematics is surely the elegant one. Unfortunately, it is almost always a distortion of the more natural thinking that led to the theorems and proofs. The necessity to establish a theorem on the basis of the axioms and previously established theorems obliges the mathematician to recast his original argument to force the theorem into the most suitable place in the logical sequence. This recast proof may be far from the original thoughts that convinced the creator his theorem was correct. Moreover, after at least one successful proof is obtained, its creator or his successors, now able to see how the essential difficulty was overcome, can usually devise a more ingenious or more direct proof. Most theorems have been reproven several times, each successive proof a remodeling of the previous one and often including generalizations or stronger results. Hence, the final theorem and proof are far from the original thoughts. Indeed, they are often shorn entirely of their intuitively grasped form. Some of the logic supplied to shore up the original intuition is entirely artificial and so trumped up and stilted as to preclude understanding. Over one hundred years ago Augustus De Morgan, one of the founders of modern logic, warned, "The student must not believe that theorems have been invented or perfected by the methods in which it is afterwards most convenient to deduce them. The march of the discoverer is generally anything but on the line on which it is afterward convenient to cut the road."

Because the deductive approach is not the understandable one, and especially because it is a distortion of the natural, intuitive approach, its value to the student is inversely proportional to its elegance. As far as the students can see, the axioms are handed down *ex cathedra* and then the logical crank is turned to grind out theorem after theorem. The students do not know where the axioms came from, why the particular ones were chosen, and where the seemingly interminable sequence of theorems is going. Even though a student may get to the point of verifying each step in a proof, he usually does not understand the rationale, the basic thought or method behind the proof. Why is this particular chain of steps used rather than some other perhaps more easily understood sequence? The student has a case. Facing such a reworked, more sophisticated, and possibly more complicated result, he cannot grasp it at all. Henri Poincaré, the leading mathematician of the late nineteenth and early twentieth centuries, makes this point: "Can one ever understand a theory if one builds it up right from the start in the definitive form that rigorous logic imposes, without some indications of the attempts which led to it? No. One does not really understand it; one cannot even retain it or one retains it only by learning it by heart."

Following a proof step by step has been compared to the way a novice at chess observes two masters play. The novice recognizes that each player makes a move that conforms to the rules, but he will not understand why a player makes a particular move rather than any one of a dozen others available to him. Nor will he perceive the overall strategy that suggests a series of moves. Similarly, watching a mathematical demonstration being made step by step, each of which is justified by some axiom or previously established theorem, does not in itself convey the plan of the demonstration, nor the wisdom of that entire method of proof as opposed to some other method. Because he may be called upon to reproduce the proof, the student is reduced to memorizing the prescribed series of steps.

Many professors, having delivered a series of theorems and deductive proofs, walk out of their classrooms very much satisfied with themselves. But the students are not satisfied. They were not involved in the real thinking and derived no stimulation from a presentation they did not understand. The learning of such proofs calls for much activity but little cerebration or celebration. The logical order of mathematical presentations is about as helpful pedagogically as the alphabetic order of words in a literary work. Colleges and universities state in their catalogues that their first objective is to encourage students to think for themselves; yet professors' presentations promote not the emancipation but the enslavement of minds.

Mathematicians have a naive idea of pedagogy. They believe that if they state a series of concepts, theorems, and proofs correctly and clearly, and with plenty of symbols, they must necessarily be understood. This is like an American speaking English loudly to a Russian who does not know English, in the belief that his increased volume will ensure understanding.

The deductive presentation of mathematics is psychologically damaging because it leads students to believe that mathematics is created by geniuses who start with axioms and reason directly and flawlessly to the theorems. Given this impression of elevated, far-ranging minds, the student feels humbled and even depressed about his own capacities, especially when the obliging professor presents the material as though he too is genius in action.

The logical or deductive approach does not convey understanding. As Galileo put it, "Logic, it appears to me, teaches us to test the conclusiveness of an argument already discovered and completed, but I do not believe that it teaches us to discover correct arguments and demonstrations." The logical formulation does dress up an intuitive understanding, but it conceals the flesh and blood. It is like the clothes that make the woman, or make the man want to make the woman, but are not the woman. Logic may be a standard and an obligation of mathematics, but it is not the essence. Nevertheless, the deductive approach to mathematics is almost universally adopted by professors.

The major reason for its popularity is that is is easier to teach. The entire body of material is laid out in a complete, ready-made sequence and all the teacher has to do is to repeat it. On the other hand, to know the intuitive meaning of a concept or proof, to penetrate to the basic idea of a proof, and to know why one proof is preferable to another call for depth of understanding. Even if the professor does acquire this understanding, it is far more difficult to impart it to students. Many teachers complain that students, particularly engineers, wish to be told only how to perform the processes they are asked to learn and then want to hand back the processes. But the teachers who offer only the logical presentation because it avoids the real techniques of teaching—leading students to participate in a constructive process, explaining the reasons for proceeding one way rather than another, and finding convincing arguments—are more reprehensible.

Since the logical approach to mathematics does not

convey understanding and even distorts the original thinking, should it be presented at all? The answer is affirmative, and the reasons have already been at least implied. Proof is a check on our intuition. It also refines and sharpens the intuition, much as argumentation with an adversary on, say, a political issue often reveals defects in our thinking.

Whether or not some liberal arts courses do much with axiomatics per se, all do offer as a prime example of "superb" mathematics the logical development of the real number system. Real numbers include the positive and negative whole numbers, fractions, and the irrational numbers such as $\sqrt{2}$, $\sqrt{3}$, π, and the like. Some history here is relevant. These types of numbers were introduced into mathematics on an intuitive and pragmatic basis. Thus, mathematicians learned to add ½ and ¼ to obtain ¾ because one-half of a pie and one-quarter of a pie amount to three-quarters of a pie. Mathematicians worked successfully with these various numbers for over five thousand years without much, if any, concern about precise definitions or the logical development of their properties. For purely professional reasons mathematicians decided in the late nineteenth century that a logical structure based on a clear, axiomatic foundation should be provided. Of course, the logical structure had to sanction what had already been established on empirical grounds. It proved to be highly artificial, contrived, and complicated. From a logical standpoint, irrational numbers in particular are intellectual monsters, and most who study this apparent aberration wryly appreciate the mathematical term "irrational." Pascal's maxim, "Reason is the slow and tortuous process by which those who do not understand the truth arrive at it," is most appropriately applied to the logical development of the real number system.

Many teachers might retort that the college student has already learned the intuitive facts about the number system and is ready for the appreciation of the deductive version, which exemplifies mathematics. If the student really understands the number system intuitively the logical development will not only not enhance his understanding, it will destroy it. As an example of mathematical structure no poorer choice could be made, because the construction is highly contrived. The development not only stultifies the mind but obscures the real ideas. Yet this topic has become the chief one in college mathematics courses. One may well conjecture that some teachers enjoy presenting the intuitively familiar facts about the number system in the recondite axiomatic approach because they understand the simple, underlying mathematics and yet can appear to be presenting profound material.

To defend teaching the logical development of the real number system many professors extol it as an example of how mathematics builds models for the solution of real problems. This is not the place to discuss applied mathematics (see the next chapter) but it is very clear that the professors who make such a statement haven't the least idea of how mathematics is applied. The example is absurd on many accounts. Let us note two. The real number system had been in use since about 3000 B.C., roughly over five thousand years before the logical "model" for it was constructed. Fortunately, no one waited for the availability of this model to apply real numbers. Nor would anyone use the model today, because the artificial, logically complex construction is as far removed from reality as heaven from earth. No one would ever think of using it even to predict anything about real numbers, let alone for a physical application. The reason for constructing the logical foundation of the real number system had nothing to do with real problems. A few professors may be aware of these facts but perhaps introduce the word "model" in this context because it has another, more pleasing association. But the logical structure in question lacks flesh and blood.

No liberal arts course is considered complete without symbolic logic. This topic, which presents the ordinary principles of reasoning in symbolic form, supposedly teaches reasoning; actually, it is farcical for this purpose. To know what symbolism to use and how to manipulate it, one must already know the common meanings of "and," "or," "not," and "implies." But the students do not have these clearly in mind, and symbolic logic only conceals them under meaningless symbols. How ridiculous to teach symbolic logic to students who still confuse "All A is B" with "All B is A." This topic is of significance only to specialists in the foundations of mathematics. Boolean algebra, which is closely related to symbolic logic, is frequently included in liberal arts courses because it can be applied to the design of switching circuits, and presumably the liberal arts students in question are going to be electronics engineers. To the students the mention of this application may well suggest switching courses.

The liberal arts courses purport to teach the power of mathematics, and they do this by teaching abstract structures such as groups, rings, and fields. A group, for example, is *any* collection of objects, such as the positive and negative integers, and an operation that performed on any two members produces a member of the collection. The operation, like the addition process applied to the positive and negative integers, must possess certain additional properties that are the abstract analogues of the familiar properties: $3 + (4 + 5) = (3 + 4) + 5$; there is a zero; and to each integer, 2 say, there is another, -2, whose sum is zero. (See also Chapter 3.)

Abstraction is indeed a valuable feature of mathematics. It reveals properties common to many concrete structures just as knowledge of the structure of mammals teaches us much about hundreds of varieties of mammals. Moreover, one who knows the abstraction can often see at once that it applies to a totally new phenomenon. Abstractions do lay bare the logical structure of several kindred concrete systems, but they are an impoverishment of the concrete as surely as the bone structure of the human body fails to present the whole man. But, some mathematicians rejoin that after presenting the abstract structure they give concrete examples. However, the concrete cases must be thoroughly understood before one introduces the unifying abstraction. To introduce as examples concrete material not yet familiar to the student is of no help in making the abstract concept clearer. In every case learning proceeds from the concrete to the abstract and not vice versa. To see the forest by means of the trees is pedagogically the only sound approach.

Many teachers favor an abstract theme, such as group theory, because they believe it to be an efficient way of imparting much knowledge in one swoop. They are under the impression that if a student is taught group theory he will automatically learn the properties of the rational, real, and complex numbers, matrices, congruences, transformations, and other topics. But a student who learns only the abstract group theory could not on this basis add fractions.

Abstractions do relate and unify many seemingly unrelated developments. However, for young people who possess little background, nothing is unified and illuminated by the abstraction. The abstract structures in question are so remote from their mathematical experiences that the values such structures grant to mathematics are no more evident than the power of philosophy to run a spaceship. To the student the abstractions are shadows that can be perceived only dimly and induce a feeling of mystification and even apprehension. Just as the human body struggles for breath in a rarefied atmosphere, so the mind strains to grasp abstractions. They may pervade the teaching but they evade the student.

What is the major problem facing this nation today? Is it inflation? Unemployment? The absorption of minorities? Women's rights? Retaining the respect of other nations? If one were to judge by the contents of the liberal arts courses, it is the Koenigsberg bridge problem. As we have related (Chapter 1), some two hundred years ago the citizens of the village of Koenigsberg in East Prussia amused themselves by trying to cross seven nearby bridges in succession without recrossing any one. The problem attracted Leonhard Euler, certainly the greatest eighteenth-century mathematician, and he soon showed that the attempt was impossible. But mathematicians will not let the dead rest in peace, and they revive the problem as though it were the most momentous one facing our civilization.

No worse a collection of dull, remote, useless, or sophisticated topics could have been chosen for a liberal arts course. Many of these topics come from the foundations of mathematics, where only specialized and professional needs justified their creation. With a few exceptions they are late nineteenth- and twentieth-century products that came long after most of the greatest mathematics was created. The best mathematicians of the past—Archimedes, Descartes, Newton, Leibniz, Euler, and Gauss—used almost none of them, for the simple reason that they didn't exist. And even the great mathematicians of the present use most of them only in specialized foundational studies. A liberal arts course that includes such topics must devote a great deal of time to convincing students that they should learn what the entire mathematical world did not miss for thousands of years, and what very few mathematicians need even today. The topics have about as much value as learning to dig for clams has for people who live in a desert. (See also Chapter 10.)

A common alternative to the melange of topics such as set theory, axiomatics, symbolic logic, and a rigorous treatment of the number system is a presentation of technical mathematics that starts about where the high school courses leave off and covers more advanced techniques. This type of course continues an old tradition. The colleges used to require that all students take more algebra and trigonometry, a requirement reminiscent of the treatment doctors of the Middle Ages prescribed for all illnesses. They "cured" every illness by bleeding the patient. Just as the cure relieved the suffering of those who bled to death, the modern technique course drains students of any vestige of respect for mathematics.

Another alternative, very popular today, is known as finite mathematics. Just what is finite about it, except perhaps the students' attention to it, is not clear. It does not include any calculus, but it does use real numbers, complex numbers, and algebraic processes and theorems that involve infinity in several ways. The content, like that of the typical liberal arts course, is a conglomeration of topics having little relationship to each other and little significance for the students to whom it is addressed. This hodgepodge of topics, set theory, symbolic logic, probability, matrices, linear programming, and game theory (we shall not here undertake to explore the nature of these topics) is one of the fads that constantly sweep through mathematics education.

Some such course, if the topics are properly chosen, might be useful to social science students. Presumably then it would contain applications to the social sciences. On examination one finds a mathematical system that describes the marriage rules of a primitive Polynesian society. To select the proper topics the organizers of the course would have to take time to find out what is really useful to the social scientists, but mathematics professors do not do this. Finite mathematics is not just a fad; it is a fraud. In any case it is not a liberal arts course.

The liberal arts courses described above give a low return on the investment of the hard work called for. Students learn brick-laying instead of architecture and color-mixing instead of painting. If these courses exhibit the liberal arts values of mathematics, then certainly student disregard and contempt for mathematics are justified.

Since all this material and its purported values fail to win over students, some professors concentrate on those who, for whatever reason, accept and pursue the material that is taught. With unconscious immodesty the professors label these students keen or bright. The "less intelligent" ones, quantitatively about 98 percent of the total, do want to know why they should learn seemingly useless material. But these professors do not recognize that they have an obligation to teach all the students.

No matter what the choice of topics, a fundamental objection to the usual liberal arts course in mathematics is that all the topics are devoted to mathematics proper. And mathematics *proper* is remote, unworldly, and even otherworldly. This thought was expressed by one of the greatest mathematicians of recent times, Hermann Weyl:

> One may say that mathematics talks about things which are of no concern at all to man. Mathematics has the inhuman quality of starlight, brilliant and sharp, but cold. But it seems an irony of creation that man's mind knows how to handle things the better the farther removed they are from the center of his existence. Thus we are cleverest where knowledge matters least: in mathematics, especially in number theory.

Moritz Pasch, a leading mathematician of the late

nineteenth century, even contended that mathematical thought runs counter to human nature.

The unnaturalness of mathematics is attested to by history. Dozens of civilizations that have existed, some celebrated for their literature, religion, art, and music, did create practical rules of arithmetic and a mixture of correct and incorrect rules for areas and volumes of common figures; but only one—the ancient Greeks—envisioned and created mathematics as the science that establishes its conclusions by deductive reasoning. Even the Greeks regarded mathematics as only a means to an end: the understanding of the physical world. Only one other civilization, the modern European civilization, has surpassed the Greeks in depth and volume of new results, and western Europe learned mathematics at the feet of the Greeks.

The subject matter of mathematics proper cannot be very attractive to most people. It deals with abstractions and this is one of the severest limitations. A discourse on the nature of man can hardly be as rich, satisfying, and life-fulfilling as living with actual people, even though one may learn a great deal about people from the discourse. Beyond the fact that the subject matter is abstract, it is *in itself* hardly relevant to life. All of mathematics centers on number, geometrical figures, and generalizations thereof. But number and geometric description are insignificant properties of real objects. The rectangle may indeed be the shape of a piece of land or the frame of a painting, but who would accept the rectangle for the land or the painting?

Moreover, the likelihood of interesting students in the material we have described is poor, especially in view of their prior education. The student who is about to begin a college course in mathematics has met the subject to some extent in his earlier education. Unfortunately, the average student leaves these earlier experiences with limited ideas as to what mathematics is and what it has accomplished in our civilization. As far as he knows the subject matter is a series of techniques for solving problems; certainly the techniques are isolated from any easily conceivable use. The failure to see values in mathematics has generally caused the student to do poorly in it, deprecate it as worthless, and shrink from further involvement.

Many professors would argue that the content of the liberal arts course is almost irrelevant. The course teaches students precise reasoning; this usually means deductive reasoning. A course in mathematics proper may teach sharper reasoning, but the students have already had three or more years of mathematics in high school, and it would seem that whatever mental training mathematics can supply would have been supplied already.

Actually, the vaunted value of deductive reasoning is grossly exaggerated. In daily life, business, and most professions deductive reasoning is practically useless. There are no solidly grounded axioms from which one can deduce what career to pursue, whom to marry, or even whether to go to the movies. On the other hand, the distinctions that must be made in analyzing character, personality, values, and good and bad behavior are far more subtle and call for a more highly perceptive and critical faculty than anything mathematics will ever develop. Deductive reasoning is not the paradigm for the life of reason. In fact, whatever faculties equip a man to understand and judge wisely about human problems are not more widely found among mathematicians nor does the study of mathematics contribute to the acquisition of such faculties. Newton was certainly not the critical thinker when he wrote about the prophecies of Daniel. In the contention that mathematics serves to train the mind, even the students smell an aroma of professional humbug.

Moreover, deductive reasoning can be learned more

readily in many other contexts. The student who is asked to recognize that because the base angles of an isosceles triangle are equal it does not follow that equality of the base angles makes the triangle isosceles, must first learn the meaning of the terms involved. He can learn the same point from the example that good cars are expensive does not imply that all expensive cars are good.

Actually, most mathematics courses do not teach reasoning of any kind. Students are so baffled by the material that they are obliged to memorize in order to pass examinations. Perhaps the best evidence for this assertion is supplied by the professors themselves. When asked why they do not allow students to use books in examinations, they usually reply, "What, then, can I ask in the examination?"

When the defense of mathematics as training in reasoning is deflated, the professors fall back on the aesthetic satisfactions mathematics offers. Though portions of mathematics are beautiful and could be presented in a liberal arts course, there are two limitations. The first is that mathematics lacks the emotional appeal of painting, sculpture, and music. The second is that students who are repelled by the mathematics they are compelled to learn in elementary and high school will not feel moved to pursue the subject to reach the few beautiful themes. The attempt to sell the beauty of mathematics to liberal arts students is doomed to fail.

Many professors proclaim that the goal of a liberal arts course should be to teach what mathematics is and what mathematicians do. No more effective means of driving students away from mathematics has ever been devised. *What is mathematics?* It is a collection of abstractions remote from life. *What do mathematicians do?* They strive for personal success, and to arrive at it some even cheat in all sorts of ways, including neglecting the interests of the very students they say they want to attract. But mathematicians create. Do these courses then teach the fumbling, the guessing, the blundering, the mental struggles, the testing of hypotheses, the frustrations, the false proofs, the insights, and other acts of the creative process? No. They teach precise definition, theorem, and proof as though God inspired the elect to proceed directly to the finished product.

Intellectual challenge and thrill of accomplishment are other values cited for mathematics. Mathematicians respond to intellectual challenge much as businessmen do to the excitement of making money. They enjoy the fascination of the quest, the sense of adventure, the thrill of discovery, the satisfaction of mastering difficulties, the pride and glory of achievement—or, if one wishes, the exaltation of the ego and the intoxication of success. Such values are present in mathematics more than in any other subject because it offers sharp, clear problems. But to obtain these values one must be interested in the subject and have already acquired some facility in it. An occasional, exceptional student persevering on his own or fortunate enough to have had one or two fine teachers may come to enjoy these values of mathematics. But such students are exceptions and are almost certainly not to be found among the students taking the liberal arts course. Moreover, people's intellects make their own claims about what they find challenging. Many find more significant challenge in law and economics.

What should a college course in mathematics for liberal arts students offer? The answer is contained in the question. The liberal arts values of mathematics are to be found primarily in what mathematics contributes to other branches of our culture. Mathematics is the key to our understanding of the physical world; it has given man the conviction that he can continue to fathom the secrets of nature; and it has given him power over nature. We now understand, for example, the motions of the planets and of electrons in atoms, the structure of matter, and the behavior of electricity, light, radio waves, and sound. And we can use this knowledge in man's behalf. Some uses of this knowledge are familiar to all of us: the telephone, the phonograph, radio and television are achievements of mathematics. Mathematics, especially through statistics and probability, is becoming increasingly valuable in the social sciences and in biological and medical research. The search for truth in philosophy or the social sciences cannot be discussed without involving the role that mathematics has played in that quest. Painting and music have been influenced by mathematics. Much of our literature is permeated with themes treating the implications of mathematical achievements in science and technology. Indeed, it is impossible to understand some writers and poets unless one is familiar with mathematical influences to which they are reacting. Religious doctrines and beliefs have been dramatically altered in the light of what mathematics has revealed about our universe. In fact, the entire intellectual atmosphere, the *Zeitgeist*, has been determined by mathematical achievements. These are the liberal arts values of mathematics and should constitute the essence of a liberal arts course.

Though this is not the place even to sketch the contents of such a course, a few elaborations may help to make clear what it can offer. The average person thinks of science rather than mathematics as providing the explanation of natural phenomena. Yet mathematics is the essence of science. Let us consider an example.

The force of gravity is involved in all phenomena of motion. The action of gravity presumably explains why the planets and their moons keep to their periodic paths and why space ships can be sent to the moon. In all motions on earth— as we walk along a level road or up or down a hill, ride in an automobile or airplane, rise from a sitting position or sit down; in the whirring of machinery; and even in the flow of blood in our bodies—the action of gravity is involved. Presumably an understanding of the force of gravity would clarify all these motions. One might argue that the mathematical law that describes quantitatively the action of gravity is useful, but that gravity is a physical phenomenon. However, to emphasize the physical force of gravity and to regard the mathematical law as an aid in analyzing and predicting the physical action is to miss the main point.

How does the sun's gravitational attraction keep the planets in their appointed paths? Is there a steel cable stretching from the sun to the earth that keeps the earth from flying off into space and confines it to its elliptical path? We have no idea of how gravity acts physically. In fact, there is no force of gravity. As Russell Baker once remarked, "You can't buy it any place and store it away for a gravityless day." It is a fiction introduced to supply some intuitive understanding of the various motions we perceive and undertake. How, then, has science been able to treat gravity, to make such precise predictions of eclipses of the sun and moon, and even to send men to the moon? The answer is that the *mathematical* law of gravitation is all that we know about this force. By means of the mathematical law and deductions from it, we can describe and predict the behavior of thousands of objects. In fact, one of Newton's great accomplishments was to show that this very law applies to both terrestrial and celestial motions.

Mathematics, then, is not only a key to our understanding of motion, it is the only knowledge we have. The same can be said of light, radio waves, television waves, X rays, and in fact all of the waves of what is called the electromagnetic spectrum. The student who enjoys radio reception of music, from Beethoven to the Beatles, should bless mathematics.

Physical science has reached the curious state in which the firm essence of its best theories is entirely mathematical, whereas the physical content is vague, incomplete, and in some cases, self-contradictory. This science has become a collection of mathematical theories adorned or cluttered with a few physical observables. To use Alexander Pope's words, the mighty maze is not without a plan, and the plan is mathematical.

In fact, it is not hard to maintain that our knowledge of the entire physical world must reduce to mathematics. As Sir James Jeans has put it, "All the pictures which science draws of Nature, and which alone seem capable of according with observational facts, are mathematical pictures." More than that, the pictures are made by man. There is no *known* physical, objective universe. We, not God, are the lawgivers of the universe.

These examples of the liberal arts values of mathematics have admittedly been drawn from the physical rather than the social sciences. The theory and predictions that mathematics supplies in the former field do describe what takes place. In the latter we have models of what could happen but doesn't; however, the social sciences are young.

Liberal arts values are so numerous and so monumental that only another sample or two can be presented here.* In the sixteenth and seventeenth centuries, thanks primarily to the work of Copernicus and Kepler, astronomical theory was converted from geocentric to heliocentric, primarily because the mathematics of the latter was simpler. Under the older view the earth was the center of the universe, and since man was obviously the most important creature on earth, man's life, goals, and activities were the most important concerns. If there is a God—and prior to the

*   For a fuller exposition see the author's *Mathematics in Western Culture.* This is a paid advertisement.

seventeenth century no one professed to doubt it—He certainly would be concerned about humans and had evidently designed the world to favor and further man's interests. But the heliocentric theory shattered all such beliefs. The earth became just one of many planets, all revolving about the sun, and man just one of many insignificant creatures on earth. How, then, at least on the basis of astronomical theory, could one believe in a God concerned with a mite, a speck of dust in a vast universe? As Matthew Arnold put it:

> The Sea of Faith
> Was once, too, at the full, and round earth's shore
> Lay like the folds of a bright girdle furled,
> But now I only hear
> Its melancholy, long withdrawing roar...
> For the world, which seems
> To lie before us like a land of dreams,
> So various, so beautiful, so new,
> Hath really neither joy, nor love, nor light,
> Nor certitude, nor peace, nor help from pain...

An even more momentous theme, which certainly belongs in any liberal arts course, is the pursuit of truth. From prehistoric times onward man has sought truths, whether through religion, philosophy, science, or mathematics. Beginning with Greek times, the one universally accepted body of truths was mathematics. The significance of this fact extended far beyond mathematics. The acquisition of some truths gave man the evidence that he could acquire them; and it gave him the courage and confidence to seek them in political science, economics, ethics, and the arts. But the creation of non-Euclidean geometry shattered centuries of confidence in man's intellectual potential. Mathematics was revealed to be not a body of truths but a man-made, approximate account of natural phenomena, subject to change and having only pragmatic sanction.

Though mathematics is a product of cultures, it in turn fashions cultures, notably our own. Just as the meaning of good literature lies beyond the collection of words on paper, so the true significance of mathematics consists in what it accomplishes for our society, civilization, and culture. In particular, mathematics is man's strongest bridge between himself and the external world. It is the garment in which we clothe the unknown so that we may recognize some of its aspects, and it is the means by which, to use Descartes' words, we have become the possessors and masters of nature. Mathematics proper may be a monument to human inventiveness and ingenuity, but it is not in itself an insight into reality. Only insofar as it aids in understanding reality is it important. And this is what we must teach.

Thus the prime goal of a true liberal arts course should not be mastery of purely mathematical concepts or techniques, but an appreciation of the role of mathematics in influencing and even determining Western culture. Appreciation, as well as skill, has long been recognized as an objective in literature, art, and music. It is equally justifiable as an objective in mathematics. The cultural aspects of mathematics, rather than the narrow viewpoint of the specialist, should be stressed in the liberal arts course to achieve an intimate communion with the main currents of thought in other fields.

Is it surprising that mathematics and the other major branches of our culture are inextricably involved with each other? Knowledge is a whole and mathematics is part of that whole. However, the whole is not the sum of its parts. The present procedure in the liberal arts course is to teach mathematics as a subject unto itself and somehow expect the student who takes only one college course in the subject to see its significance for the general realm of knowledge. This is like giving him an incomplete set of pieces of a jig-saw puzzle and expecting him to put the puzzle together. Liberal arts mathematics must be taught in the context of human knowledge and culture.

Professors must learn that mathematics proper is not the most important subject for the nonprofessional. Even some of the best professional mathematicians did not grant the subject supreme importance. Newton regarded religion as more vital and said that he could justify much of the drudgery in his scientific work only on the ground that it served to reveal God's handiwork. But of course Newton was just a lowly physicist. Gauss ranked ethics and religion above mathematics, but Gauss, too, was as much a physicist and astronomer as a mathematician. Weyl's words—it is an irony of creation that man is most successful where knowledge matters least, in mathematics—bear repetition.

Why do mathematics professors teach pointless material to liberal arts students and ignore the truly cultural values of mathematics? The sad fact is that most professors are themselves ignorant of these values. Some may be powerful engines of mathematical creativity but limited to their tracks. They work in their own mental grooves and naively assume that what they value is eminently suitable. If the thought of including, say, applications to science should occur, they would banish it because they know they might be embarrassed by the students' questions. The elitist, narcissistic mathematician who presents his own values, curiosities, and trick problems is totally unfit to be a teacher in any course.

The charge that most mathematics professors are culturally narrow may seem incredible. But one must remember that mathematics is in large part a technical subject in which one can be highly proficient as a cabinetmaker is proficient among carpenters; and one would not necessarily be surprised to learn that a cabinetmaker is neither cultured nor a pedagogue. Competent researchers need not and generally do not know the broader values of mathematics. Certainly most do not care about the art of pedagogy. These professors present to all students just those values that they as professional mathematicians see in their subject—and they do not question their own values. They wish to have students appreciate the abstractions, the rigorous reasoning, the logical structure, the crystalline purity of mathematical concepts, and the presumed beauty of proofs and result. Non-Euclidean geometry, the most dramatic and shocking event in recent intellectual history, is to them just another topic.

The damage done by such professors extends beyond what they inflict on students. Because they are so involved in their own research, they ignore and even disparage anything outside their own specialty. This attitude discourages young professors, many of whom, less indoctrinated, do recognize the need for a truly liberal arts course but refrain from any action for fear that the older men will regard them as trafficking in trivialities. They are made to feel that any talk about music or philosophy is a default on their real obligation to teach mathematics, and in fact to train future mathematicians. Hence, departure from the norm is discouraged.

It is tragic that professors teaching in a liberal arts college, which is purportedly devoted to educating the whole person and to instilling interests and attitudes, are not themselves interested in learning material closely related to their own subject.

C.P. Snow, in his famous lecture, *The Two Cultures and the Scientific Revolution*, deplored the gulf separating the scientists and humanists. The former, self-impoverished, disdain the humanistic culture and even take pride in their ignorance of it. The latter respond by wishful thinking to the effect that science is not part of culture but rather mechanization of the real world. They wish neither to understand nor to sympathize with the nature and goals of the scientific enterprise, and they pity the scientist who does not recognize a major work of literature. Snow's more severe criticism is directed toward the humanists who pretend that their cultural interests are the whole of culture and that the exploration of nature is of no consequence. Actually, it would have been more just to rail at the scientists, particularly the mathematicians. One cannot expect them to make exposition of their subject their mission in life; but where they have the chance, even the obligation, to offer a liberal arts course to about two million freshmen each year and thereby make the meaning and significance of their subject clear to themselves and the students, they default.

The students in a liberal arts course, many of whom are the most intelligent of our youth, are our best bets for producing broadly educated men and women who would be unhampered by petty parochialism and fully alive to the interrelationships of not just two cultures but of all knowledge. Some might even become members of a now rare breed of politicians who have some idea of what scientists are doing. Surely the mathematics course should aim at such goals.

Mathematicians are digging their own graves. Student protests for relevant education, though not always wisely formulated, are fully justified in the case of mathematics. A course in college mathematics is fast disappearing as a general requirement, and mathematicians will lose not only jobs but also any significant role in the liberal arts college. The mere fact that mathematics has been taught for centuries certainly is no assurance that it will be retained. Both Latin and Greek are the languages of the cultures that have contributed most to the fashioning of Western culture, and Latin was the international language of educated people until about one hundred years ago. But both languages have practically disappeared from modern education. Indeed, even the study of classical Greek and Roman culture has practically disappeared.The danger that mathematics, too, may be dropped from the liberal arts curriculum has already occurred to many mathematicians, and some are growing concerned about the public image of their subject. But apparently they are not sufficiently aroused to utilize the best medium already at their disposal to improve that image.

Socrates was condemned to death for corrupting the morals of the youth of Athens. What punishment should be meted out to professors who degrade mathematics, cause the students to hate the subject or intensify the already existing hate, and in many cases poison the students' minds against all learning? And should the universities be exculpated for entrusting the education of liberal arts students to specialists or graduate students who are the living refutation of liberally educated people?

7

The Undefiled Mathematician

The most vitally characteristic fact about mathematics is, in my opinion, the quite peculiar relationship to the natural sciences, or, more generally, to any science which interprets experience on a higher than a purely descriptive level.

John von Neumann

Until about one hundred years ago all mathematicians would have accepted von Neumann's conception of mathematics. Certainly the three men whom mathematicians nominate as the greatest of all time—Archimedes, Newton, and Gauss—did more scientific than mathematical work and in fact justified their mathematical investigations by mentioning or describing the applications that warranted the mathematical research. Even those whom many cite as pure mathematicians—Carl Jacobi, Karl Weierstrass, and Bernhard Riemann—not only applied mathematics, but pursued mathematics proper to clarify, rigorize, and extend the theory and technique they already knew to be applicable. This is not to deny that some of these men engaged in subjects, such as the theory of numbers, whose attraction is primarily aesthetic value or intellectual challenge; but one need only count the years they devoted to purely aesthetic subjects as opposed to those bearing on science to determine which research they regarded as more important.

The situation is quite different today. Though the greatest mathematicians of recent times—Hermann Weyl, David Hilbert, Felix Klein, and Henri Poincaré—would have endorsed von Neumann's characterization of mathematics, now only about one out of ten mathematicians devotes himself to problems of the physical and social sciences, and many of these are employed in governmental and industrial laboratories. Among professors, about 5 percent do applied work. The rest are totally ignorant of science and do not undertake any problems bearing on it. The days when mathematicians saw the hand of God in the motion of the planets and the stars are gone. Especially in view of the facts that science and technology have expanded at least as much as mathematics and that social and biological problems are now being tackled mathematically, this reversal calls for an explanation.

In part we have already given it. Professors are expected to publish. The older applied fields—mechanics, elasticity, hydrodynamics, and electromagnetic theory—have been explored for one, two, or three centuries and the outstanding problems are no longer simple. The newer applied fields—quantum mechanics, magnetohydrodynamics, solid state physics, meteorology, physical chemistry, and molecular physics—presuppose an extensive background in physics. As for the social and biological sciences, these are more complicated, and so far successes have defied the best brains. What, then, should professors, especially young people who have yet to earn rank and tenure, publish? The obvious answer is to pick some specialty in pure mathematics and to invent problems that can be solved. Since the editors of the journals come from the same milieu as the professors, this artificial research is as readily publishable today as are the most profound papers of mathematical science.

What has this alteration in the nature of research to do with pedagogy? The answer is that most mathematics professors no longer teach either the uses of mathematics in science nor how to apply mathematics to scientific problems. Perhaps the best example of the detachment of mathematics from science is furnished by the teaching of calculus. This subject is the crux of applied mathematics, and next to the liberal arts course it is the one that serves the most students. Prospective engineers, physical and social scientists, actuaries, technicians, and medical and dental students take it to learn how to apply the subject. How then do the professors teach calculus?

During the first four decades of this century the calculus course was a series of mathematical techniques taught mechanically and imitated by the students. The students worked but didn't have to think. Of course, this type of course was neither very helpful nor enlightening to the students, but in view of the level of mathematical knowledge in the United States it was about as good as could be expected. As more students flocked to college, and as a college education became a prerequisite for a good job or a professional career, the meaninglessness of the calculus course became more apparent. Though the professors had become more knowledgeable in mathematics proper, they were not prepared to teach a calculus course suited to the interests and needs of the students. The subsequent squirming and twisting reveal the modern professor's evasion of his obligations.

Until about 1945 mathematics students took analytic geometry before calculus. Analytic geometry deals with a new and vital idea, the coordination of curve and equation, an idea that is used extensively in calculus. To "improve" the calculus course mathematics professors decided to start students with calculus and to teach the requisite analytic geometry as it was needed. Analytic geometry consequently got short shrift. This consolidation also meant asking the student to learn two major techniques simultaneously. Moreover, since the study of analytic geometry obliges students to utilize algebra and trigonometry, when they took analytics before calculus they were better prepared for the necessary uses of these tools in calculus. Subordinating analytic geometry to a topic in the calculus course deprived students of a sorely needed background. Most, therefore, did poorly.

The professors "saw" the remedy for this trouble. They decided to incorporate more algebra and trigonometry, along with the analytic geometry, in the calculus course. This move proved to be still more disastrous, and the professors backtracked. Now the algebra, trigonometry, and analytic geometry are packed into a one- or two-semester course that is called precalculus—a nice semantic device to avoid admission of the original error.

Some professors took another tack. In addition to including algebra, trigonometry, and analytic geometry, they adopted a rigorous approach to calculus. That is, they included the theory as well as the technique. This move, they evidently believed, would make the calculus course understandable. But the theory of the calculus is highly sophisticated. The best mathematicians, from Newton and Leibniz, who worked in the late seventeenth century, to Cauchy, who worked in the early nineteenth century, struggled to understand the logical foundation of the calculus, and Cauchy was the first to make the proper start. A sound foundation was not achieved until Karl Weierstrass, fifty years after Cauchy's breakthrough, cleared up the mess. Certainly, then, the theory is not easy for beginners to grasp. One may be sure that the very same teachers who believe that students beginning calculus can absorb a theoretical foundation would have been swamped, in their own student days, by such a presentation. Nevertheless, having finally grasped the theory after some years of study, they forgot their own experiences and acquired a missionary zeal to spread the light.

The proper pedagogical approach to any new subject should always be intuitive. The strictly logical foundation is an artificial reconstruction of what the mind grasps through pictures, physical evidence, induction from special cases, and sheer trial and error. The theory of the calculus is about as helpful in understanding that subject as the theory of chemical combustion is in understanding how to drive an automobile. This approach through theory, which had its heyday in the mid-1960s, had to be abandoned.

The prewar teaching of pure technique was not successful; the inclusion of algebra, trigonometry, and analytics was no more so; and the inclusion of rigor was soon found to be a disaster. What measure could the mathematicians adopt? Since the precalculus course, when instituted, took care of the elementary material needed for calculus, the obvious move was to include material beyond calculus. Bits of linear algebra, vector analysis, differential equations, and other topics (some of which have no relevance to calculus) were therefore included in the calculus course, which is now a hodgepodge of topics and a mélange of unrelated techniques. (See also Chapter 10 on texts.)

In all these maneuvers the professors have avoided the one measure that would make the calculus course meaningful and serve the purpose for which it is intended—namely, to make it the introduction to applied mathematics. Mathematics proper and calculus especially are mazes of symbols and manipulations of symbols. As such, these have no meaning or purpose. They are the shadows of substance and have as much meaning as the notes of a musical score to one who cannot hear the composition they describe. The symbols have no life, but properly interpreted they can tell us about the vital forces that affect almost every aspect of our lives. Only the applications supply meaning and motivation. Since most calculus students will be engineers or physical scientists and intend to use calculus, what better insight could they be given than examples of where calculus achieves results?

Calculus offers excellent opportunities not only to apply mathematics but also to show how physical arguments suggest deep mathematical results. For example, we know that a ball thrown into the air rises and then falls. At the highest point in its path the velocity must be zero, else it would continue to rise. Because the velocity is the rate of change of distance with respect to time, what this physical happening suggests is that the rate of change of one variable with respect to another must be zero at the maximum value of the first variable. This is a basic theorem of calculus.

There is, however, an obstacle to the introduction of physical problems that might supply motivation, meaning, and application: Most mathematics professors know no science and will not extend themselves to learn it. Those few who know enough to present the simplest physical applications fear the questions that may ensue.

Many professors realize that calculus proper is dull and meaningless but, not prepared to offer real applications, they put on a pretense of doing so. They assign the following typical problem: Find the velocity of an object that moves with an acceleration of $a = 5t^3 - 3t^2 + 4t - 16$. To find velocity knowing the acceleration is indeed an application of calculus, but what object in this universe moves with the acceleration stated? Perhaps a drunken driver. Some professors are more realistic. It happens that objects moving near the surface of the earth, if one neglects the resistance of air, are subject to a downward acceleration of thirty-two feet per second each second. Hence, it does make some sense to pose problems of motion involving this acceleration. However, such problems are simple and do not exhibit the power of the calculus. Motion in a vacuum should be followed by more realistic problems involving motion in an atmosphere. The parachutist who could not rely upon the resistance of air would not, after one drop, have to rely upon anything. Perhaps the professors who confine their applications to motion in a vacuum are preparing students for life on the moon, which has no atmosphere, while convincing them that life on earth is intolerable.

Calculus texts offer other "real" applications. A man six feet tall is walking away from a street light at the rate of five feet per second; the problem asks how fast the man's shadow is lengthening when he is ten feet away from the light. The problem deals only with the shadow of reality.

Professors also introduce problems in which physical terms such as "center of gravity" and "moment of inertia" are used. But the physical meanings of these concepts and their uses are not taught. The consequence is that the gravity of these problems produces moments and even hours of inertia in the students.

Of course, curiosity might induce some people to solve any problem. But curiosity not only kills cats; it kills interest in mathematics courses that pose pointless problems. To teach calculus without real applications is to ask people to sit down at a table set for a dinner where no food is served, or to teach grammar but never to mention literature.

Why are professors content to teach artificial, dull and pointless applications? Such problems have been in calculus texts for fifty or more years. The professors learned how to solve them when they were students. Why bother to dig up new and more significant ones and incur much more work in learning to present them if there is no pressure to improve the course? Surely it is boring to repeat the same deadly material year after year; but then all of teaching is a chore to be disposed of as quickly as possible.

The deficiencies in the calculus course are exemplary of a glaring deficiency in the entire mathematics program, graduate and undergraduate. Though the major reason that students take mathematics is to use it, only a few undergraduate and graduate schools offer applied mathematics. There are courses and texts that are titled "Applied Mathematics," but these are pitiful. They offer mathematics that *can* be applied, for example differential equations, but at best they mention where the topics are applied. They omit the problem of analyzing physical phenomena to determine which factors or features can be neglected and which must be incorporated in the mathematical formulation; and they fail to teach at all the process of translating physical facts into mathematical language. Since no physical problems are treated, the payoff—what one learns through the mathematics about physical phenomena—is missing. Moreover, most of the texts contain tidbits from various mathematical areas used in applications but no one topic is pursued in depth. If these texts make any impression on the students it is only to bewilder them. They are shifted from topic to topic so quickly that nothing sticks. One is reminded of the whirlwind tours of Europe which cover ten countries in ten days. The tourists return home uncertain as to whether the Eiffel tower is in Paris or Prague.

Mathematicians are, of course, aware of the existence and importance of applied mathematics, and they are sensitive to the charge that they are neglecting it. They justify their purely mathematical offerings on the ground that they are teaching students how to build models for the solution of real problems. As new physical problems are tackled, presumably all the applied mathematician or scientist will have to do is run through his files and select the model that fits his problem. But this type of model-building is a waste of time, and advocacy of purely arbitrary mathematical creations reveals ignorance of what applied mathematics involves.

God may have designed the world mathematically but evidently He did not intend to make that design readily accessible. Mathematics is not emblazoned on the face of nature. Several crucial and difficult steps are necessary to mathematize genuine physical problems. Any real situation contains dozens of elements whose relevance must be considered. If one is studying the motion of a ball, the color can surely be neglected, but the shape and size may not be negligible. On the other hand, if one is studying the reflection of light from some surface or the transmission of light through some translucent material, the color of the surface or the material may be critical. In the study of the motion of a planet around the sun, both the planet and the sun may be regarded as point-masses, that is, the mass of each can be regarded as concentrated at one point. The reason is simply that their sizes are small compared to the distance between them. On the other hand, if one is studying the motion of the moon around the earth, the size and shape of the earth must be taken into account. Should the attraction of both bodies by the sun also be taken into account? That depends upon the problem to be solved. To predict the tides of the oceans on the earth, the sun's attraction does matter. However, to study the precession of the earth's axis, that is, the change in the direction of the imaginary line through the North and South Poles, the sun's attraction can be ignored. The more exact the answer required, the more care must be exercised to be sure that the relevant factors are taken into account. Simplification of a problem by discarding the irrelevant factors is a crucial step, and it presupposes an insight that may, of course, be deepened by experience.

After simplifying a problem one must apply physical principles. (In studying the motion of the earth around the sun, the law of gravitation is a fundamental principle.) Such principles are usually supplied by physicists, but the translation of the physical principles and other relevant information about the particular problem into the language and concepts of mathematics must be done by mathematicians. The concepts may not be available and may have to be created. In fact, precisely the need to treat problems of motion motivated the creation of the calculus. New concepts are added constantly as new problems are tackled.

Once the mathematical formulation is achieved, the next stage is the solution of a mathematical problem. There are times when the applied mathematician is lucky. The mathematical problem may have been solved in the course of some earlier study. And one may be sure, if this does prove to be the case, that the solution was originally sought in behalf of a real problem. More often, unfortunately, the mathematical problem is a new one that calls for original work.

Other problems and processes, such as approximation adequate for the use to which the solution is to be put, enter into applied mathematics, but the major point is that mathematical models cannot be constructed a priori and then called upon when needed. One cannot prefabricate useful models. The mathematics involved in real problems is far too complex and special to be conjured up by the free play of the imagination. A physical problem comes to the hands of mathematicians as a rock encrusted with sediment and mud. It is up to the mathematician to remove the dross,

chip away the encrustations, polish the rock, and bring forth
ultimately a blazing gem of physical truth. There is an art of
applying mathematics and an art of teaching that art.

One cannot expect students to solve new applied
problems. But they will be required to do it in their
professional work; therefore, they should be taught all that is
involved in the entire process. To delude students into
believing that the study of solely mathematical structures
and processes suffices is to falsify the account. The
professors' contention that the study of mathematical
models prepares students for applying mathematics is, at
best, wishful thinking to rationalize their own ignorance
and, at worst, conscious deceit.

When challenged that the values of mathematics proper
do not mean much to potential users, many professors retort
that students will learn the applications in other courses. But
to ask students to take seriously theorems and techniques
whose worth will be apparent one, two, or several years later
is a grievous pedagogical error. Such an assurance does not
stir up incentive and interest and does not supply meaning to
subject matter. As Alfred North Whitehead has advised,
whatever value attaches to a subject must be evoked here
and now.

Still another argument offered by professors against the
teaching of applications is that it imposes a heavy burden on
the students; they must learn the mathematics and the
relevant physics, say, and they must also learn to relate the
two. But this argument is specious. Carefully chosen
applications do not require much extramathematical
background, and the little that is required can readily be
included in the mathematics course. Moreover, the teaching
of mathematics is expedited by tying it in with applications.
These provide motivation, which mathematics proper does
not. Equally important, the only meaning the concepts had

for the mathematicians who created them and the only meaning students will find in courses such as calculus derive from physical or, more generally, real situations.

Professors also use the argument that they cannot cover the syllabus if they include real applications. But even if this argument has force, and it does have some, in what sense is the ground covered? The professors cover the topics in the syllabus but the students are buried so deeply under an landslide of ideas and techniques that they no longer see light. The ground is covered over the students.

In view of the fact that the application of mathematics to science and engineering is its most vital and widespread use, the absence of applied mathematical courses is as deplorable as the absence of honesty in our political leaders. But since most mathematicians are no longer capable of offering applications, they shun them as infections in a sound body. The intransigence of mathematics departments in meeting their obligations to students majoring in areas such as science and engineering is notorious. They teach as though mathematics is all we know and all we need to know. Unfortunately, it is in the most prestigious universities that the professors are allowed to take the position that they are authorities unto themselves. They are autonomous and teach what they will. Syllabi for courses and a planned sequence of courses, essential in a cumulative subject such as mathematics, are detested and ignored by the high and mighty. Professors often palm off the teaching of courses addressed to science and engineering students on the younger members of the faculty or on graduate students, who have no choice of the courses they must teach.

What subject matter do professors teach? Except in a course such as calculus, where the content is prescribed, they favor their own specialties. These in turn are determined by their research. It is not surprising to learn that courses in mathematical logic, abstract algebra, topology, the theory of numbers, functional analysis, and axiomatics dominate the undergraduate and graduate curricula. The technical nature of these subjects need not be examined. What matters is that these subjects, very fashionable today, constitute a one-sided account of mathematics. All are pure; that is, devoid of applications. Professors prefer virginity to bedding with science.

Professors are specialists and they tend to view the world of mathematics through the medium of their own specialty. Certainly there should be courses for prospective pure mathematicians. But all mathematicians should be informed about the chief value of mathematics—namely, its interplay with science and the amazing fruitfulness of that interplay. Here man demonstrates the magical power of his mind, and how mathematics bridges the gap between his mind and the real world. Or, as Max M. Schiffer, professor of mathematics at Stanford University, has pointed out:

> The miracle of mathematics is that paper work can be related to the world we live in. With pen or pencil we can hitch a pair of scales to a star and weigh the moon. Such possibilities give applied mathematics its vital fascination. Can any subject give the would-be mathematician—initially at least—a stronger and more natural interest? And what about the non-mathematician? Deny him introduction to this subject, and his appreciation of our cultural heritage must inevitably be inadequate. For mathematics in the broadest sense is instrumental not only to our understanding, but also to our changing the world we live in.

Mathematics majors may be free later to pursue any branch of pure mathematics, but not to know the chief role of their subject is to be ignorant, no matter how many research papers they may write later. A goodly number of the courses should deal with applications to science, and some physics should be part of the education of every mathematician. Further, since the prospective mathematicians are most likely to become university teachers, they should be prepared to teach prospective scientists and engineers.

That the typical mathematics professor was not required to learn any science can be charged to the graduate school professors, who (with few exceptions) are a collection of specialists in various areas of pure mathematics. These professors, who face the task of educating scientists, engineers, and future teachers of such students and refuse to do so, are irresponsible. Were they in an industrial or commercial organization, they would be fired. That such dismissals do not take place in the universities is due only to the fact that mathematicians as a group are nearly all in the same position, and not even the chairman, who is one of the group, would wish to take action. Many professors claim that they take pride in their work and do fulfill their obligations as teachers. But their actions belie their words.

What is wrong, then, is not that professors do not know what they are teaching but that they do not know how to relate their subject to the rest of knowledge and to life. Education is evaded within academic walls as well as without, and it is more often the professors rather than the students who do so. Professors may not be consciously dishonest or aware of their pedagogical ineptness. However, the lack of clear standards of teaching, the professors' ignorance of pedagogy, and the obligation, which many take too literally, to cover ground prescribed in syllabi produce the same effect as incompetence and dishonesty. The practices of the past are not re-examined, and the challenges that true education might pose are ignored.

The student who goes to college to prepare for a career is

certainly justified in doing so, and the colleges have been implicitly accepting such a goal while explicitly talking about liberal education. Hence, courses for business-oriented students, statisticians, actuaries, and scientists must be given the attention other courses get. True, vocational or professional interests clash to some extent with purely academic interests, but the former should not be submerged or ignored. Nevertheless, myopic professors impose their own interests on the students, with the result that their courses are largely useless to most students and to society. Students are accused of resistance to intellectuality, but their resistance is to arrogant and indifferent professors who, in the name of academic freedom, serve themselves.

Mathematicians have abandoned science in an age whose major achievements are scientific and most of whose principal problems will be solved largely by resort to science. They live in a self-imposed exile from the real world. Blinded by a century of ever purer mathematics, the professors have lost the will and the skill to read the book of nature. Like the mathematicians Gulliver met in his voyage to Laputa, they live on an island suspended in the air and leave to others the problems of earthly society.

The extent to which they have abandoned science is perhaps best indicated by the words of Marshall Stone, who served over the last few decades as a professor at Yale, Harvard, and Chicago:

Nevertheless the fact is that mathematics can equally well be treated as a game which has to be played with meaningless pieces according to purely formal and essentially arbitrary rules, but which become intrinsically interesting because there is such a great fascination in discovering and exploiting the complex patterns of play permitted by the rules. Mathematicians increasingly tend to approach their subject in a spirit

which reflects this point of view concerning it. . . . I wish
to emphasize especially that it has become necessary to
teach mathematics in a new spirit consonant with the
spirit which inspires and infuses the work of the
modern mathematician, whether he be concerned with
mathematics in and for itself, or with mathematics as an
instrument for understanding the world in which we
live. . . . In fact, the construction of mathematical
models for various fragments of the real world, which is
the most essential business of the applied mathematician, is nothing but an exercise in axiomatics. . . . When
an acceptable modern curriculum has been shaped in
terms of its mathematical content, one must still be
concerned with the spirit which animates the subject
and the manner in which it is taught. It is here that it is
highly appropriate to demand that, even in the earliest
stages, an effort should be made to bring out both the
unity and abstractness of mathematics.

Stone's words have not gone unchallenged. Professor
Richard Courant, formerly head of the pre-Hitlerian world
center for mathematics at the University of Göttingen and
more recently the founder and head of what is now called
the Courant Institute of Mathematical Sciences of New York
University, has denounced this abrogation of the essence of
mathematics:

A serious threat to the very life of science is implied in
the assertion that mathematics is nothing but a system
of conclusions drawn from the definitions and postu-
lates that must be consistent but otherwise created by
the free will of the mathematician. If this description
were accurate, mathematics would not attract any
intelligent person. It would be a game with definitions,
rules and syllogisms without motive or goal. The notion
that the intellect can create meaningful postulational
systems at its whim is a deceptive half-truth. Only
under the discipline of responsibility to the organic

whole, only guided by intrinsic necessity, can the free mind achieve results of scientific value.

The various sins of pedagogy were attacked again by Courant:

> Perhaps the most serious threat of one-sidedness is to education. Inspired teaching by broadly informed, educated teachers is more than ever an overwhelming need for our society. True, curricula are important; but the cry for reform must not be allowed to cover the erosion of substance, the propaganda for uninspiring abstraction, the isolation of mathematics, the abandonment of the ideals of the Socratic method for the methods of catechetic dogmatism.... At any rate it would be without doubt a radical and vitally needed remedy for many ills in our schools and colleges if a close interconnection between mathematics, mechanics, physics, and other sciences would be recognized as a mandatory principle which must be vigorously embraced by the coming generation of teachers. To help such a reform is a solemn obligation of every scientist.

The abandonment of applied mathematics by most mathematicians is a blow not merely to pedagogy. It is a threat to the very existence of mathematics itself. Problems which stem from the real world are the lifeblood of mathematics. It must remain a vital strand in the broad stream of science or it will become a brook that disappears in the sand. The entire history of mathematics shows that physical science has supplied the inspiration, vitality, and fruitfulness of this subject.

Nor should one overlook the value of applied mathematics to technology and thereby to humanity. That men and women now work thirty-five or forty hours a week instead of eighty, that their homes are better built and more comfortable, that they enjoy quality phonograph records and television, that they can receive medical treatments which cure diseases or at the very least prolong their lives, these and a multitude of other benefits are due in large measure to mathematics.

To speak of an applied mathematics program as though it were one of many programs or as though there were two kinds of mathematics, pure and applied (or monastic and secular, as some would put it), is a concession to the current practice. But it is a misrepresentation of mathematics and mathematics education. There is just one subject: mathematics. The chief function of that subject, and its chief claim to support by society and to an important role in education, reside in what it does to help man understand the worlds about him—physical, social, biological, and psychological. So far the successes have been mainly in the physical sciences, but judged by the time scale of civilizations, mathematics is young.

Many mathematicians today and of recent years would dispute this evaluation. Mathematics, they say, is what mathematicians do, and since most mathematicians have no interest in anything but the subject itself the relationship to other fields is irrelevant. No doubt many are sincere. Since they do not know the magnificent and powerful uses of mathematics they can be indifferent to them. But they should heed more of von Neumann's words:

> As a mathematical discipline travels far from its empirical source, or still more, if it is a second and third generation only indirectly inspired by ideas coming from "reality," it is beset with very grave dangers. It becomes more and more purely aestheticizing, more and more purely *l'art pour l'art*. This need not be bad, if the field is surrounded by correlated subjects, which still have closer empirical connections, or if the

discipline is under the influence of men with an exceptionally well-developed taste. But there is a grave danger that . . . at a great distance from its empirical source, or after much "abstract" inbreeding a mathematical subject is in danger of degeneration. . . . In any event when this stage is reached the only remedy seems to me to be the rejuvenating return to the source: the reinjection of more or less directly empirical ideas. I am convinced that this was a necessary condition to conserve the freshness and vitality of the subject and that this will remain equally true in the future.

Does the abandonment of science by most mathematicians mean that science will be deprived of mathematics? Not entirely. The Newtons, Laplaces, and Hamiltons of the future will create the mathematics they need, just as they did in the past. Formerly such men, though they were honored as mathematicians and served as mathematics professors, were basically physicists. In today's world they are cast out by the mathematicians but they find their place in science departments. Nevertheless, the services of many able people are now lost to science and to the necessary pedagogy.

The failure of mathematics departments to cater to science, engineering, and social science students has had the expected effect. The physical science, social science, and engineering departments of many universities offer their own mathematics courses. In some institutions, statistics and probability courses are given in half a dozen departments. In almost all colleges and universities computer science has broken away from mathematics and is a separate department. Clearly, the users of mathematics have decided that mathematics is too important to be left to the mathematicians. Should this practice expand, it will be the end of mathematics education as such. Curiously, though some mathematics professors resent this competition for students and the takeover of what they regard as their province, others exhibit a remarkable "broadmindedness." Their reaction is, Well, now we don't have to teach the useful mathematics; we can teach what we please.

Though the teaching of mathematics courses by other departments and schools such as engineering may seem to resolve the problem of giving students the kinds of mathematics courses that would be more useful to them, it is not the realistic solution. A relatively minor objection is that it leads to endless duplication of courses, which the universities can ill afford. More consequential is the fact that mathematics, physics, economics, and the various branches of engineering are now vast domains and a professor is hard put to it to master even a portion of any one of these domains. A physics professor, for example, may know selected portions of mathematics, but he cannot teach mathematics effectively. He does well if he can teach successfully his area of physics. If, on the other hand, these several users of mathematics were to hire mathematicians to teach the mathematics they want taught, they may not require the services of a full-time professor and may settle for a graduate student or adjunct. However, even if a number of full-time professors are employed but are segregated in nonmathematical departments, they will be isolated and will fail to benefit from contact and communication with colleagues who can stimulate and educate each other. The departmental organization of disciplines has drawbacks, but it is the best we have. Mathematics is far too important to be taught in dribs and drabs.

The neglect of applied courses and the poor teaching by mathematicians has had the consequence that more than half of the students who enter college as science majors either drop out or turn to other fields. About 50 percent of the engineering students do not graduate. Other factors enter, of course, but poor mathematics education is certainly a major cause. The courses are lethal weapons that snuff out the intellectual interests and sometimes the academic lives of the students.

One who peruses the current mathematical journals will find articles advocating the institution of applied mathematics programs in the undergraduate and graduate schools, and may therefore infer that mathematicians are now genuinely concerned with this obligation. But closer investigation will reveal that this sudden interest is caused not by any recognition of student needs or guilt about the poor courses that have been offered but is, rather, selfishly motivated. Academic positions for Ph.D.'s in mathematics are now far fewer than the number of Ph.D.'s being turned out by the graduate schools. However, industry and the federal government do employ applied mathematicians. Hence, graduate students and Ph.D.'s are being urged to prepare for the available jobs. Were they to accept this recommendation, most would have great difficulty in acquiring the proper training.

If one were to judge mathematicians by their pedagogy, one could take either of two positions, depending upon whether he was charitably or critically disposed toward them. Under the former, they would be characterized as so introspective as to be unworldly, deeply in love with their subject, intensely concerned with the progress of mathematics, and so far superior to ordinary mentalities that they cannot appreciate normal problems. Under the second view, their preference for their own values and interests marks them as selfish, overly absorbed in themselves, culturally narrow, and indifferent to society's interests and to the needs of students. In either case the mistakes made by professors bury their pupils just as surely as the mistakes of doctors bury their patients.

Nevertheless, the professors who teach pure, abstract mathematics and maintain that they are successful do provide some lift to all of us. We can be quite sure once again that there is a heaven and that there are angels. Certainly the students must be angels to be able to absorb what the professors say they teach them, and the professors evidently derive their sustenance from heaven because they have their heads in the clouds.

8

The Misdirection of High School Education

Great God! I'd rather be
A pagan suckled in a creed outworn;
So might I, ...
Have glimpses that would make me less forlorn;

William Wordsworth

In a world in which changes take place with bewildering and discomfiting rapidity, mathematics education has provided one fixed landmark: the high school mathematics curriculum. This program is not only fixed; it is, thanks to the university ideology and practices that determine it, rigid and seemingly immovable. But the high school curriculum is certainly outmoded. Let us examine it.

The ninth grade is devoted to algebra and the subject matter has been almost entirely a series of disconnected, atomized processes, a mélange of topics: factoring, operations with polynomials such as x^2+5x+b, operations with fractions such as $(2x+5)/(3x+7)$, laws of exponents, and the solution of equations of various degrees and with one or more unknowns. This approach to mathematics proper is most unfortunate for, on the whole, algebra is a means rather

161

than an end. It is the spelling, grammar, and rhetoric of most mathematics, but it is not literature. If students of English were asked to spend years on preparation for reading—which is what the mathematics student does in eight years of arithmetic and the one year of algebra—without ever being introduced to the pleasures of reading, how would they react?

Beyond the fact that algebra per se is a means rather than an end, it involves a specific difficulty that bothers almost everyone—the use of letters. Just as number symbols are a hurdle for elementary school children, so literal symbols are a hurdle for high school students. To speak of $3x = 5$ is not quite so terrifying as to speak of $ax = b$, wherein a and b are any numbers. In principle the idea is simple. Instead of talking about John and Mary, algebra talks about men and women. But generalizations about men and women are meaningful only to those who have had lots of experience with individual men and women. The analogue for algebra is lots of experience with various manipulations of numbers. Unfortunately, few students have more than a nodding acquaintance with arithmetic at the time they are propelled into algebra.

The use of literal coefficients did not occur to mathematicians until about 1600 A.D., roughly two thousand years after first-class mathematics had been produced by the Greeks. This fact in itself should warn teachers that students will balk at letters and that special devices are needed to make the transition from numbers to letters acceptable.

Whereas algebra proper is both meaningless and point-less, geometry, which students learn in the tenth grade, does have intuitive meaning. But the pointlessness is still glaring. Here, many dozens of theorems are proved in a logical sequence and, to the students, the goal seems to be to prove as many as possible. They are confined to learning inflexible and obtuse trains of thought that permit no derailment. Just why anyone wants these theorems and how these theorems and their proofs were ever conceived is not treated.

The eleventh grade curriculum repeats the ninth grade subject matter because students did not learn that material, adds more algebraic processes, and introduces trigonometry—in which, among other topics, the students learn many identities at the risk of losing their own.

The twelfth grade contents, taken by relatively few students, has not been stable. But the material, whether solid geometry or the beginnings of calculus, is taught dogmatically and has been no more enlightening or enthralling than the preceding subjects.

How did this curriculum come about? It was fashioned in the nineteenth century by professors. In fact, as we have already seen (Chapter 2), the subjects that are now taught in the high schools were, in the United States, first taught only in the colleges. They were gradually moved down into the high school curriculum, but retained essentially the content that had been taught at the college level. During the years of transition, and even long afterward, committee after committee re-examined the curriculum. In detail the recommendations made by these committees did differ; some would teach factoring before exponents, and others the reverse. But in essence all agreed upon this traditional curriculum. The arguments they accepted in favor of it are hackneyed. The value of mental discipline was never questioned, and this purported value—that mathematics teaches thinking—freed the committees from any real concern whether the content was right. In addition, utility in daily life, preparation for college (after the subjects were transferred to the high school), and the learning of the higher truths in the noblest branch of our culture were also offered to justify the status quo. There seemed to be no doubt that

$a^2 - b^2 = (a + b) (a - b)$ would uplift any soul.

There has been one change in recent years. The ineffectiveness of the traditional education, made more evident by the needs of World War II, stimulated some university professors to devote themselves to the reform of the curriculum. But these professors were even less informed about high school education than their predecessors of the last century or the early part of this century. Hence the reformers, too, thought in terms of the values usually claimed for mathematics education, such as preparation for the mathematics to be taken in college. Their contribution was to impose rigor, generalization, abstraction, and an emphasis on structure onto the traditional content. All the arguments marshaled against such features in preceding chapters apply with greater force to the high school curriculum, because the students have yet to find out what mathematics is all about.

Consider, for example, structure. Structure of what? The answer seems to be structure of abstract mathematical systems. It is true that the collection of positive and negative whole numbers and fractions—that is, the rational numbers—do have some properties that the whole numbers alone do not have. And the real numbers have the same structure as the rational numbers. Since these structures are all that high school students know, and not too clearly at that, what can the study of structure mean? One might as well ask students who have yet to see a dog to learn the structures of various species of animals. Moreover, what does structure mean in Euclidean geometry and trigonometry? The word is fashionable, but certainly not applicable to high school mathematics.

The new curriculum did not significantly alter the basic high school material. It secured the rigidity of the curriculum by adding rigor to the proofs and by burrowing deeper into the foundations through set theory and symbolic logic. (See Chapter 6.) These "innovations" may have ensured the stability of the several mathematical edifices, but they also pulled the students so far down into the dark earth that they could no longer see the surface. Where light and air were needed, the new mathematics added steel beams. This reform, far from being an improvement, is more accurately described as a disaster. Fortunately, the new math is a passing aberration and there is no point to beating dead bones.

To the mathematician all of this standard material, whether in traditional or new math guise, is self-justifying; to the student it is self-condemning. The contents of all four years, certainly as presented, is abstract, dull, boring, and intrinsically meaningless. It consists at best, to use Alfred North Whitehead's phrase, of inert ideas. That students should have difficulty with it is inevitable. Psychologists found years ago that human beings could not remember more than six or so nonsense syllables one hour after they were memorized. If students haven't already acquired a dislike for mathematics in elementary school, the high school mathematics, particularly algebra, will surely engender it. The aftermath of traditional mathematics is revulsion. It is incredible that mathematicians and pedagogues could ever have believed and still do believe that this curriculum has any value. To be sure, one or two students in some classes do take to mathematics, either because they like it or because they wish to guarantee admission to college, but mathematicians would be the first to caution us that one swallow does not make a summer.

Let us review the arguments presented in defense of the traditional curriculum. One argument has been that this material is useful in later life. Does the average educated person use this knowledge in daily life? Do even mathematics teachers, who know the subject, ever use the quadratic formula, the Pythagorean theorem, or the trigonometric identities outside of the classroom? The honest answer is no. Of course, the students learn that a straight line is the shortest distance between two points. That's useful. But even a donkey knows that: Put some food at a distance from him and watch the path he takes. The subject matter of high school mathematics—that is, the subject matter per se—is worthless knowledge. One or two topics, such as the calculation of compound interest, may prove useful. But the exceptions do not alter the general assessment.

And the high school courses have not taught thinking, pedagogues' assertions to the contrary. The traditional algebra teaches memorization of processes. Geometry purportedly emphasizes proof and therefore thinking; however, because the proofs are arranged in a logical sequence and this sequence is not natural, the rationale eludes the students and they are forced to memorize here, too. The students' appreciation of proof in geometry is epitomized in their oft-repeated remark that geometry is where you make proofs in two columns.

There is another frequently advanced argument for teaching mathematics: The subject is beautiful. Whereas in the college liberal arts course topics can be chosen for beauty, what is taught on the high school level is not notable for its aesthetic value. The choice of topics was based on what is needed for further education in mathematics. All the preaching and rhapsodizing will not make such ugly ducklings as factoring, adding fractions, and the quadratic formula attractive. Even the fact that the sum of the angles of a triangle is 180° is hardly attractive; a sum of 200° would at least be a round number.

Moreover, beauty is a matter of taste, and in the case of mathematics the appreciation of beauty calls for a certain sophistication. It is in fact fortunate for society that not too many people are attracted by the esoteric beauty in mathematics. Our world would be sadly affected if even 10 percent of American high school students should decide that they wished to pursue mathematics for mathematics' sake. One good medical doctor is worth a thousand mathematicians who pursue mathematics for its beauty.

Omitted thus far is another value that champions of traditional mathematics cite: intellectual challenge. But should we sacrifice the 99 percent to the 1 percent who respond to this particular intellectual challenge?

The clinching argument that advocates of the traditional curriculum have used endlessly is that students need high school mathematics to go to college. It is inexplicable how anyone could take such a position in 1900, or even in 1930 when no more than 25 percent of the high school graduates went to college. Though the percentage has increased steadily, it is still true that no more than 50 percent of those who graduate from high school enter college. Moreover, at least 25 percent of those who start high school do not graduate.

Of those who go to college, at most 10 percent will need the techniques and theorems now taught in the high schools. The others either take no mathematics in college—many colleges no longer require a mathematics course for the bachelor's degree—or take a liberal arts course that does not use most of the currently taught high school mathematics. Do the colleges require the conventional high school mathematics for admission? They have not demanded it for many years. Some do not ask for any high school mathematics. Almost all of the others will accept any respectable two-year curriculum. Hence, one cannot justify the present high school curriculum on the ground that it prepares students for college.

The argument concerning preparation for college has also been self-defeating. Faced with the dull traditional curriculum, fewer students have been taking academic mathematics. In 1928, when about three million students attended secondary school 27 percent took ninth-grade algebra and 18 percent took plane geometry. In 1934, the corresponding percentages for five and one-half million students had dropped to 19 and 12. Since 1934 the number of students attending secondary school has increased sharply, but the percentage taking academic mathematics has declined still more. Those who still take it do so only because most colleges, as already noted, require some courses, though the precise content is not restricted to the traditional subject matter.

Moreover, how appealing is the motivation—preparation for college—to the students? What normal fourteen-year-old really knows whether he is college bound? His parents may have definite plans, but these may go awry for many reasons. One certainly will be the child's reaction to high school work, and the traditional mathematics courses are not likely to induce a favorable reaction. Even at the time of graduation from high school, many youngsters are still uncertain about whether they will go to college. They often decide to do so only because of external pressures or circumstances. They may imitate what friends do or follow the path of least resistance.

To all of the arguments for the traditional high school mathematics curriculum the advocates of the new math added that our country's need for mathematically sophisticated manpower required the secondary school curriculum to bring students more quickly to the frontiers of research in pure and applied mathematics. In view of the sequential nature of mathematics, such an objective is preposterous.

Whether or not the values claimed for mathematics are

indeed present, certainly student readiness and capacity to appreciate these values should have entered into the considerations of the framers of the curriculum. The results of their failure to consider effectiveness have been notoriously poor. If we may judge by the results on the Scholastic Aptitude Tests, the performance of the best group—the college-bound—has been getting poorer year by year. More students are going on to college, but the high schools should be succeeding with all of these, at least. Additional evidence of ineffectiveness is the fact that remedial mathematics is now the biggest problem that the colleges face. In the two-year colleges 40 percent of the students taking mathematics are enrolled in remedial courses.

One cannot dismiss the traditional material without taking into account that some high school students will eventually use mathematics professionally. Since students at the high school stage do not know whether or not they will need mathematics later, everyone, so the argument goes, should be required to take it. However, only about 5 percent of the high school students will ever use mathematics and their needs should not dominate. They can be met in a manner that will be suggested later.

Is it, then, a mistake to require mathematics of all academic high school students? Not at all. The arguments against the teaching of mathematics are against the kind of mathematics that has been taught and against the justifications traditionally and currently given. At least in the United States, which is the only society that has sought to teach mathematics to all students, the vital material and the reasons for teaching it have been ignored entirely. The framers and reformers of the curriculum have approached the problem from the wrong direction. The values and objectives have been those that professors enjoy or respect.

Before considering an alternative approach, let us note a critical distinction between elementary and high school mathematics education. Everyone has to know some arithmetic and a few elementary geometrical facts merely to get along in life. Arithmetic has basic practical value, much as reading and writing have. However, as we have already pointed out, this is not true of algebra, geometry, and the other high school subjects. Whereas we are obliged to present the elementary school material, we are freer to choose the material to be taught on the high school level. What are the objectives of an academic high school education? The courses in literature, history, science, economics, and foreign languages are intended to enable future adults to live more insightfully, wisely, and enjoyably. In short, they are an introduction to our culture. The mathematics courses should serve the same objectives. Hence, the mathematics we teach should be worth knowing for the rest of the lives of *all* students. It should contribute to a truly liberal arts education wherein students get to know not only what the subject is about but also what role it plays in our culture and our society. We must teach not just what mathematics is but what it does.

Teachers will undoubtedly ask, "Where and what is the material that constitutes a liberal arts curriculum?" To an extent we answered this question when we considered the contents of a freshman liberal arts course (Chapter 6). Admittedly, the intellectual level of the high school program and the background that it presumes must be taken into account. But there is a mass of suitable material, much of which can be culled from diverse sources. A few examples may convince doubting Thomases.

Elementary algebra deals with simple functions such as $y = 5x$, $y = 3x^2+6x$, and the like. As purely mathematical expressions they are dull, devoid of interest; but they can be and are used to represent an enormous variety of motions, the motions of balls, projectiles, rockets, and spaceships. Motion on the moon, which exhibits striking contrasts with motion on the earth, provides an exciting theme. The motion of objects dropped into water is another readily understandable and interesting phenomenon. Further exploration of simple functions leads to the law of gravitation, and to remarkable calculations such as the mass of the earth and the mass of the sun. Functions are not merely symbolic expressions; they are laws of the universe, and they encompass the behavior of grains of sand and the most distant stars.

Algebra can also be applied to the study of elementary statistics and probability. These techniques are used to obtain vital knowledge about the distribution of height, weight, intelligence, mortality, income, and other facts of interest and concern to every would-be educated person. The efficacy of medical treatments; the control of quality in production; and the prediction of future prices of commodities, population growth, and genetic traits such as susceptibility to diseases are achieved with the same tools. The reliability of conclusions reached on the basis of statistics and probability should also be taught. The well-known quip, "There are lies, damned lies, and statistics," should certainly be taken seriously in the study of statistics, because so many of the conclusions that are hurled at us are not supported by the data.

The uses of high school geometry are manifold. The determination of the size of the earth, the distances of the planets from the sun, and the periods of the planets; the explanation of eclipses of the sun and moon; and the calculation of inaccessible lengths, such as the height of a building, the width of a canyon, or the distance across a lake—all, though utilizing only the simplest geometry, are remarkable feats. Equally accessible through geometry is the behavior of light. If a light ray travels from A to B (Figure 1) via reflection in a mirror, the path it takes, namely, the one for which angle 1 equals angle 2, is the shortest one it could possibly take. Thus the actual path APB is shorter than, for

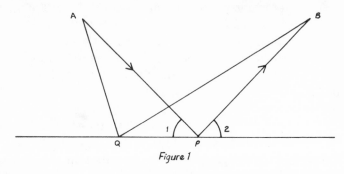

Figure 1

example, AQB. Moreover, since light travels in this situation with constant velocity, the shortest path requires the least time. In fact, in practically all situations light takes the path requiring the least time. Now light is surely inanimate. How, then, does it know to choose that path? Why does it seek "efficiency"? Here one can touch upon one of the grandest doctrines that man has proposed and in some centuries unquestionably affirmed. There is an order in nature, a design behind each phenomenon, and mathematical laws reveal and express that design. Who instituted that design? Perhaps best left unanswered, this question can be raised at the high school level.

A beginning can be made toward teaching the power of mathematical abstraction. What the pure geometry of the reflection of light says is that if one wants to go from point A to line QP and then to B, the shortest path is the one for which angle 1 equals angle 2. Suppose now that QP is a railroad track and a station is to be built to serve the people of towns

A and *B*. Where should the station be situated so that the total distance the people of *A* and *B* must travel to reach the station is a minimum? We have the answer. It is *P*. Suppose instead that a school or a telephone central is to be located along *QP* to serve towns *A* and *B*. Again, minimum distance or minimum telephone lines require that the school or central be placed at *P*. Thus, one geometrical theorem answers many different practical questions.

Other uses of elementary geometry can deal with a description of the functioning of our solar system, geography, perspective in painting, the structure of atoms and molecules, architecture, elementary engineering, surveying, navigation, and even clothes design.

If this cultural approach is extended to the third year of high school mathematics, wherein trigonometry is usually taught, many more values of mathematics can be demonstrated. When light travels from one medium to another, as from air to water, it changes direction; that is, it is refracted. The law of refraction can be simply stated in trigonometric terms. The importance of knowing this law is readily demonstrated because it is used in designing eyeglasses, and in understanding the functioning of the human eye and treating its diseases. The trigonometric functions enable man to analyze the pitch and quality of musical sounds, and this analysis is used in designing the telephone, radio, phonograph, and television. Of course, none of the above-mentioned applications calls for a knowledge of set theory. But teachers should be willing to make some sacrifices in behalf of 100 percent of the students.

That mathematicians of the past were inspired by real problems and found the meaning of mathematics through them is beyond dispute. Equally beyond dispute is that applications to real problems are a pedagogical necessity. However, there is some question about which applications will be interesting and meaningful to students. Only experience will enable teachers to determine the best choices. Fortunately, the choices are so numerous and so varied, and interest in the real world is so much more widespread than interest in abstract mathematics, that attractive applications can surely be found. At the very least, it is easier to arouse curiosity about real problems than about mathematics.

Though authority can be found to defend almost any position, the words of some of the most famous educators of the past warrant reading. Friedrich Wilhelm Froebel (1782-1852), the founder of the kindergarten, stated:

> Mathematical demonstrations came like halting messengers.... On the other hand, my attention was riveted by the study of gravitation, of force, of weight, which were living things to me, because of their evident relation to actual facts.

Aristotle asserted that there is nothing in the intellect that was not first in the senses. The use of real and, especially, physical problems serves not only to answer the question of what value mathematics has but also gives meaning to it. Negative numbers are not just inverses to positive integers under addition; they are the number of degrees below zero on a thermometer. The ellipse is not just a peculiar curve; it is the path of planets and comets. Functions are not sets of ordered pairs; they are relationships between real variables such as the height and time of flight of a ball thrown into the air, the distance of a planet from the sun at various times of the year, and the population of a country over some period of years. To rob the concepts of their meaning is to keep the branches of a fruit tree and throw away the fruit.

There is another value to be derived from developing mathematics from real situations. One of the greatest difficulties that students encounter in mathematics is solving verbal problems. They do not know how to translate the verbal information into mathematical form. Under the usual sterile presentations in the traditional and new mathematics curricula, this difficulty is to be expected. On the other hand, if mathematics is not just applied to the real world but is drawn from the real world, as happened historically, its applicability is no longer a mysterious process. The difficulty students now experience in applying mathematics to real problems is very much the difficulty a French youth, who knows only French, would have in translating his thoughts into English. If, however, he is brought up in a bilingual region as are, for example, the French Canadians, he can certainly express himself in both languages.

The inclusion of applications offers still another advantage. As we have noted, most of mathematics arose in response to the desire or need to solve real problems. As the teacher treats such problems, he can include an account of the historical background in which the problems were tackled and even include some biography of the men involved. If skillfully interwoven, rather than added as disconnected appendages, the history and biography will not only enliven the courses but also will teach an equally important lesson: Mathematics is produced by human beings responding to a variety of problems. It is a living body of ideas that developed over the centuries and grows continually.

The defenders of the traditional material often rejoin that they do teach applications. Let us look at just a few of them. There are work problems such as the ditch digger's dilemma: "One man can dig a ditch in two days and another in three days. How much time will be required if both men dig it together?" Such problems create pointless work. Then there are tank-filling problems for students who have no swimming pools to fill. And there are the mixture problems: "How many quarts of milk with 10 percent cream and how many quarts of milk with 5 percent cream must be mixed to make one hundred quarts of milk with 50 percent cream?" Such problems are perhaps most useful to farmers who wish to fake the cream content of their milk.

And we shouldn't neglect to mention the time, rate, and distance problems, such as up- and downriver travel, for students whose desire to go anywhere except out of the classroom will not be aroused.

Some authors of algebra texts point to "truly physical" problems. For example, Ohm's law states that the voltage E equals the current I times the resistance R. In symbols $E = IR$. Calculate E if $I = 20$ and $R = 30$. But if the concepts and the use of the law are not explained to the students, the current doesn't drive any mental motors.

The proposal that the first two or three years of secondary mathematics contain all sorts of genuine applications may seem radical, but a little perspective may correct this impression. In the seventeenth century mathematics courses comprised astronomy, music, surveying, measurement, perspective drawing, the design of optical instruments, architecture, and the design of fortifications and machines. In the intervening centuries some of these topics lost importance and were dropped from the mathematics curriculum. The expansion of mathematics itself and of knowledge generally has compelled educators to drop other topics. But the isolation of mathematics from all applications relevant to our times cannot be tolerated, even if the inclusion of applications necessitates covering less mathematical topography.

Beyond the demands of pedagogy, we must recognize that mathematics did not develop independently of other human activities and interests. If we are compelled for practical reasons to separate learning into mathematics, science, history, and other subjects, let us recognize that this separation is artificial. Each subject is an approach to knowledge, and any mixing or overlap, where convenient and pedagogically useful, is to be welcomed.

The need to relate mathematics to our culture has been stressed by Alfred North Whitehead, one of the most profound philosophers of our age, and a man capable of the most exacting abstract thought. In his essay "The Aims of Education," written in 1912, Whitehead says:

> In scientific training, the first thing to do with an idea is to prove it. But allow me for one moment to extend the meaning of "prove"; I mean—to prove its worth.... The solution which I am urging, is to eradicate the fatal disconnection of subjects which kills the vitality of our modern curriculum. There is only one subject-matter for education, and that is Life in all its manifestations. Instead of this single unity, we offer children Algebra, from which nothing follows; Geometry, from which nothing follows.... Our course of instruction should be planned to illustrate simply a succession of ideas of obvious importance.

In another essay of 1912, "Mathematics and Liberal Education" (published in his *Essays in Science and Philosophy*), Whitehead goes further:

> Elementary mathematics ... must be purged of every element which can only be justified by reference to a more prolonged course of study....
> [T]he elements of mathematics should be treated as the study of a set of fundamental ideas, the importance of which the student can immediately appreciate; every proposition and method which cannot pass this test, however important for a more advanced study, should be ruthlessly cut out.... [S]implify the details and

emphasize the important principles and applications.

In 1912 Whitehead was addressing himself to essentially the same curriculum that we teach today. The criticisms and positive recommendations still apply and, in fact, with all the more force because the high schools now teach a more diverse group.

To present mathematics as a liberal arts subject requires a radical shift in point of view. The traditional approach presents those topics of algebra, geometry, intermediate algebra, and trigonometry that are necessary to further students' progress in mathematics per se. The new approach would present what is interesting, enlightening, and culturally significant, with the inclusion of only those concepts and techniques that will serve to further the liberal arts objective. No technique for the sake of technique should be presented in the first two or three years. In other words, the material should be objective-oriented rather than subject-oriented.

But what do we do for the future professional user of mathematics? Admittedly a small but appreciable percentage of the students will become mathematicians, physicists, chemists, engineers, social scientists, technicians, statisticians, actuaries, and other specialists whose work requires mathematics. Of course these students, too, should know the cultural significance of mathematics. Moreover, students who are already inclined toward a specific career will certainly take an interest in mathematics if they see how the subject is involved.

But if by pursuing the cultural objective we do curtail somewhat the technical preparation of those students who will need mathematics later, what can we do for them? Those who are convinced by the end of the eleventh year of schooling that they do wish to pursue mathematics for some professional use should be offered an optional technical course in the twelfth year. Because these students know that the career they intend to follow will require mathematics, they will be better motivated and, most likely, quite capable of rapid progress. They should be able in one year to acquire a considerable technical background that might well include far more than they may acquire in the present traditional curriculum. In fact, these students, mixed in at present with indifferent or poorly prepared high school students, don't go far in the ninth and tenth grades, and in the usual eleventh-grade class they are bored by repetition intended for the poorer students.

Beyond content there are many pedagogical considerations—such as the treatment of proof; getting students to enter into the discovery of results; the use of laboratory materials, among which for present purposes we may include the computer; and testing. However, these pedagogical problems lie beyond our present concern. Our discussion of content may be sufficient to indicate that all is not well in high school education.

More relevant, in view of the notoriously poor results of mathematics education, is the question of why the basic content of the high school mathematics curriculum has remained fixed. The answer is that for many generations the mathematics departments of the universities and of most four-year colleges have taught the same subject matter to all prospective high school mathematics teachers. In fact, on the whole, the prospective teachers are taught the same subject matter that is taught to all mathematics students. The professors project their own values and interests; in the present case, in which courses are directed to prospective high school teachers, student needs do not count. To study the problems of high school education and to fashion courses that would enable prospective teachers to meet those problems would call for the full-time efforts of at least one professor in each college or university department. But the archenemy of all undergraduate education—the pressure to do research—precludes such attention. Though the professors of the schools or departments of education do teach prospective teachers how to teach, they are unable to counter the domination of content by the academic departments. Hence, their impact on curriculum is nil.

Consequently, the high school teacher is limited in knowledge and restricted to goals and values set by mathematics professors and administrators. However much he may sense the poverty of the material he is teaching, he has neither the expertise nor the power to change the curriculum.

Also, unsurprisingly, curriculum reform is not often welcomed by older teachers. Indoctrinated only in traditional, isolated, unmotivated mathematics, the teacher confronted with the challenge to teach more meaningful and more purposive material shrinks back in fear. When asked, for example, to teach applications involving acceleration, he throws up his hands and argues that the concept is too difficult for the students or that it presupposes a knowledge of physics. However the students and the teachers ride in automobiles. To get a car moving and to stop it one must accelerate or decelerate. A moving automobile rarely travels at a constant speed but accelerates and decelerates constantly; hence, the notion is certainly intuitively familiar. The students, who are still young and open to ideas, would not find the concept difficult—nor would the teachers, had they not been conditioned to concentrate solely on mathematics. In fact, it is even vital to teach acceleration. It is too rapid deceleration that causes many passengers in automobiles to go through the windshield. To require teachers to teach applications does not impose any real hardship. Actually, very little knowledge of science is required to teach applications of high school mathematics; every teacher can acquire it. Prospective mathematics teachers should certainly study some science during their undergraduate days.

A rich, vital, and attractive high school curriculum can be fashioned. Reform should be led by experienced, knowledgeable, broadly educated professors. Specialists might serve as consultants but certainly should not lead. It would be equally important to have cultured nonmathematicians participate. Their judgment as to what the future citizen would find valuable should take precedence over that of the specialists.

High school mathematics that consists of algebraic techniques, proofs of geometrical theorems, and mazes of abstract concepts and symbols will continue to reduce the students to a state of bafflement and loathing. At present, many conclude that they do not have mathematical minds when, in fact, they have not had informative, inspiring, and stimulating mathematics education. Though sincere teachers have been imparting what they themselves have learned—skills and proofs—unconsciously and sometimes consciously they work hocus-pocus on their students, presenting and repeating opaque formulas, sometimes to the admiration but almost always to the bewilderment of their charges.

The current high school curriculum is a threat to the life of mathematics. If the schools do not offer a more rewarding and meaningful curriculum, then rather few students will take mathematics. Requirement for admission to college, which is the main factor keeping academic mathematics alive in the secondary schools, may not sustain high school mathematics in the future. Almost certainly, it will cease to be a requirement. If high school mathematics is not made more attractive and significant to the student, mathematics as an integral part of general education will die quietly and its soul will rise to heaven through an atmosphere of irrelevance.

9

Some Light at the Beginning of the Tunnel: Elementary Education

I had been to school . . . and could say the multiplication table up to 6 x 7 = 35, and I don't reckon I could ever get any further than that if I was to live forever. I don't take no stock in mathematics, anyway.

Mark Twain (Huckleberry Finn)

Of all the levels of mathematics education, from elementary to graduate, the elementary is the most difficult to teach. The primary reason is that we do not know enough about how young children learn. We know little more about how older children learn, but they are no longer as much dependent on the teachers.

In view of the lack of any sound knowledge of pedagogy it is not surprising that from the very founding of primary schools in the United States until about 1950, sheer ignorance determined the method of teaching arithmetic—drill in skills. Certainly, in the nineteenth century and prior to that there were no knowledgeable professors to train elementary school teachers and, in fact, most were trained in normal schools, which could hardly be considered college-level institutions. However, even when the training of elementary school teachers was taken over by the liberal arts colleges and the schools of education, no improvement was made. The liberal arts professors were really not concerned about the mathematics that should be taught to prospective elementary school teachers, and the professors in the schools of education knew no more about how to teach than did their predecessors in the normal schools.

Educational psychology was introduced early in this century as a formal discipline in the schools of education, and prospective teachers were required to take a course in this subject. But the professors had nothing to offer. They endorsed and even applauded rote learning. The leader of this school of thought, Professor Edward L. Thorndike, maintained, as we noted elsewhere (Chapter 2), that students should be taught to respond automatically to any given question or problem. Repetition of the stimulus-response would develop skills. In fact, Thorndike maintained, in *The Psychology of Arithmetic* (1924):

> Reasoning is not a radically different sort of force operating against habit but the organization and cooperation of many habits, thinking facts together. ... Almost everything in arithmetic should be taught as a habit that has connections with habits already acquired and will work in an organization with other habits to come. The use of this organized hierarchy of habits to solve novel problems is reasoning.

The value of this methodology can readily be tested. One has only to ask an educated adult, How much is 3/4 divided by 2/3? Inexplicably, teachers continued to demand rote learning in accordance with Thorndike's psychology while at the same time advocating the values of mental discipline and transfer of training in reasoning to other fields.

A new era was inaugurated in the 1950s when mathematics professors decided that the elementary- and secondary-school curricula needed reform. Arithmetic became an abstract, logically developed subject. To present this new approach teachers had to be trained or retrained. And so new courses in content were offered by mathematics departments.

Because set theory had been introduced as a topic to be taught to elementary school students, prospective teachers had to be taught set theory. The ludicrousness of what was and is, in the main, still taught may be illustrated by the attention given to the empty set, usually denoted by ϕ. Thus, if no apples are present, the empty set of apples is denoted by ϕ. Some texts for teachers ask the students to name several different empty sets. Others ask them to prove that the empty set is unique. Further, there is strictly a logical distinction between an object and set of objects. A set or collection of books is not a book. Likewise, there is a distinction between the empty set and the set containing the empty set. The latter is denoted by $[\phi]$. It "follows" that $[\phi]$ is not empty, because it contains the empty set. The inclusion of one set in another, say a set of six books in a set of nine, is denoted by the symbol $\subset$. Would it not then be "important" to ask, Is $\phi \subset [\phi]$?

More generally, these newer courses stressed the deductive structure of mathematics, which means proof made by deductive reasoning from axioms. Presumably, without such proof nothing can be asserted. But in the elementary school deductive proof should play little, if any, role. Inductive arguments, physical evidence, measurement, and pictures are the basis for learning and understanding. However, the prospective elementary school teacher, sitting at the feet of a distinguished professor, is in no position to reject or refute his teachings. Hence when the student becomes a teacher he feels like a traitor if he induces young children to accept a process or technique on the basis of, say, pictorial evidence rather than deductive proof.

If the material just described is the best that mathematics

departments can offer prospective elementary school teachers, then they should offer no courses in academic mathematics to these students.

One defense of such college courses is that teachers must prepare children for the uses that will be made of mathematics twenty and thirty years from now. The professors also argue that understanding of fundamental principles will narrow the gap between elementary and advanced knowledge. Just what these future uses and advanced mathematics have to do with the education of children is not treated.

The content, then, of the courses usually addressed to prospective elementary school teachers is decidedly poor. To make matters worse, the teaching of such courses is often entrusted to graduate students or young instructors on the ground that the subject matter is simple and does not require the mastery of mathematics that the older professors have. The real reason is that the professors don't wish to bother with courses for nonmathematicians.

The psychologists hastened to defend the new approach. Jerome Bruner's statement, in *The Process of Learning*, "We begin with the hypothesis that every subject can be taught effectively and in some intellectually honest form to any child at any stage of development," has been quoted endlessly and used—in particular by the proponents of the New Mathematics—to defend teaching abstractions. The saving feature of this statement is its vagueness. What is an intellectually honest form of teaching Immanuel Kant's *Critique of Pure Reason* to six-year-olds?

It seems fair to say that most parents can discern as much about how children learn as psychologists have taught us. That all people, children in particular, must be amply grounded in concrete experiences before they can appreciate an abstraction is readily observed. That children differ in the interest they take in learning, in the rate of learning concrete material, and in the ages at which they can grasp even simple abstractions, whether these differences are genetic or conditioned, is obvious.

With or without help from psychologists we can readily appreciate that the hardest part of any subject is the beginning, because it is strange. It is also the most important part, because the patterns of thought and the attitudes acquired at this time become established in young minds and then are hard to change. The obstacles to the learning of arithmetic are especially formidable. The first is that it deals almost immediately with abstractions. The number 5 is a meaningless symbol to most five-year-olds. More difficult is the question, Is 7 greater than 5? One might as well ask a child to calculate the distance to the star Sirius. Moreover, the abstractions become more remote from experience in the higher grade levels. Negative numbers are far more difficult to grasp and eluded the best mathematicians for over one thousand years after they were introduced.

Another obstacle to the learning of arithmetic stems from our ingenious but unnatural way of expressing quantity. Our number system uses positional notation; the two "1's" in 151 have entirely different meanings. The sophistication of positional notation is evident from the fact that of the many civilizations that developed some system of arithmetic, only one, the Babylonian, conceived that scheme. The Europeans learned the idea from the Babylonians, and improved the scheme somewhat by shifting to base ten from base sixty. Though positional notation facilitates the operations of arithmetic, it makes comprehension more difficult. The rationale of long division is not readily understood and the usual square root process is even more difficult to comprehend. The elementary school teacher is obliged, if he is to avoid mindless drill, to apportion his efforts and the available time between teaching skills and imparting understanding. Neither task is easy.

Still another difficulty is the precision of the language that must be employed. In ordinary discourse many ambiguities are tolerated and the meaning is usually deciphered from the context. Thus, a person who addresses a letter "Dear John and Mary" surely means that John and Mary are both dear to him, but grammatically the "Dear" modifies only John, or at least it can be so interpreted. Likewise, the banks that advertise that they are happy to extend small car loans are not specific as to whether they mean that they will make loans only on small cars or that loans on any cars would be small. Such ambiguities cannot be tolerated in mathematics; thus $10 - 7 + 3$ is not the same as $10 - (7 + 3)$; if the latter is intended, the parentheses are essential. Likewise, $3 \times 4 + 5$ is intended to mean $(3 \times 4) + 5$ and not $3(4 + 5)$. (Parentheses would, in fact, help; instead, elementary school students are taught that the strong operations—multiplication and division—must precede the weak operations of addition and subtraction.) Mathematical language must be precise even though it necessitates symbolism and stiltedness.

The difficulties we have described cannot be evaded. In contrast to the situation at the secondary-school level, there can be little choice of contents in elementary school mathematics. Every adult must be able to perform arithmetic and use common geometrical formulas for area, perimeter, and volume; and in our society some knowledge of statistics and probability is almost as imperative. Whether these last two topics should be taught in the elementary school or reserved for the secondary school is debatable.

In view of the content and the few principles we do possess about how children learn, what should prospective elementary school teachers be taught? A truly cultural liberal arts course, such as was described earlier (Chapter 6), would indeed be more helpful in the education of all people and certainly to teachers of elementary mathematics. To the latter it would give some perspective on the place of mathematics in human affairs. Some of the knowledge gained could even be utilized in their teaching. But the primary need of prospective elementary school teachers is re-education in arithmetic. They know no more than they learned in elementary school and it is commonly conceded that these teachers, like most adults, are insecure in that knowledge. Many even dread the period that they must devote to the subject. These failings seriously hinder effective teaching. Re-education would certainly help, particularly because as college students these prospective teachers are more motivated than they were as elementary school students. Beyond re-education, the course could give insight into arithmetic and geometry, which would fulfill the essential requirement that a teacher must know more than he teaches. Thus, whereas positional notation in bases other than ten should not be taught to elementary school students, it should be taught to prospective teachers. The usual square root process cannot be made entirely clear to children, but it can be to teachers. Terminology such as inverse, commutative, associative, and the like can be taught to prospective teachers, though it should not be imposed on youngsters. Since the elements of statistics and probability are entering the elementary school curriculum, certainly future teachers should know more about these topics than they will teach.

No doubt mathematics professors will object that such a course is not college level and will resent having to use their precious time and energy to teach such simple material. But the farce of teaching the kind of course they do give and their ignorance of the needs of elementary school teachers do not seem to trouble them. Apparently, bewildering students with endless questions on the empty set is college level.

Despite the at present poor preparation of elementary school teachers, some progress in elementary education has taken place in the last few years, especially in the lower grades, whereas none is noticeable at the other levels. One reason may be reaction to the New Math. Of course, the New Math was a disaster at both the elementary and secondary levels, but the extent of the disaster was more evident at the primary school level. When students could not add 9 and 8, almost everyone was shocked. That a high school student could not solve $x + 3 = 7$ should have been equally shocking, but since solution of this problem is not required in daily life and is beyond most adults anyway, the failure to master simple algebra was not attacked as severely.

In any case, teachers concerned with the elementary school seem at last to have learned the obvious. Abstractions, however simple, acquire meaning only in terms of a host of concrete experiences. Thus, texts of the last few years have pictures of people, animals, houses, candy, and other objects familiar to children, and the whole numbers and operations with whole numbers are discussed for a long time only in connection with these pictures. That $2 \cdot 3 = 3 \cdot 2$ is no longer based on the commutative law or taught by pure rote.

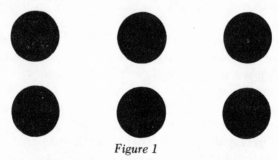

Figure 1

Children are asked to look at a rectangular array of discs (Figure 1) and they see that two rows of three discs and three columns of two discs still total six. They soon recognize that the order in which any two numbers are multiplied is immaterial and thus learn the commutative property of multiplication. Fractions are parts of pies and it is visually obvious that 2/6 of a pie is the same as 1/3 of a pie. The student then readily understands the principle that one can multiply the numerator and denominator of a fraction by the same number and not change the value of the fraction itself.

Building up arithmetic on the basis of real situations accomplishes more than the teaching of skills. If a child learns to perform the four operations but has not constantly associated them with real problems, he will be in the position of not knowing which operation to use in a given instance. If asked to give three children equal portions of 1½ pies, should he divide? If so, should he divide 3 by 1½ or 1½ by 3?

As the teaching of arithmetic progresses to the higher grades, wherein operations with large numbers, more complicated fractions, decimals, and percentage are the topics, teachers, poorly prepared in their college education, often ask, What real phenomena or situations can we employ to make the arithmetic meaningful and purposeful? Some suitable themes are the same as those suggested for the first two years of high school, but the questions raised would be simpler. For example, given the speed of sound in air, students could be asked how far sound travels in a stated number of seconds or how long it takes a sound signal to travel a given distance. These questions are not intrinsically exciting, but one can readily introduce some that are. Sound travels at about 5,500 feet per second through the earth but at only 1,100 feet per second through air. Hence, a distant sound reaches the auditor more quickly through the ground. This is why Indians put their ears to the ground to hear a rider on horseback sooner. One can ask for the difference in time if the distance is one mile. Another application of the speed of sound uses the interval of time required for a sound to reach a mountain and for the echo to return to calculate the distance to the mountain. The same principle is used to determine the depth of the ocean and of harbors, and to determine the distance of one submarine from another.

Work with the speed of light, which is 186,000 miles per second, can also lead to some novel facts. Given the distance from the earth to the sun how long does it take light from the sun to reach us? A calculation would show that it takes light from the sun eight minutes to reach the earth. Having learned this, students might be asked, Do we see the sun where it is now? Of course not. We see where it was eight minutes ago. It may have exploded in the meantime.

Speed, time, and distance can be applied to problems of sports. For example, a line drive—that is, a ball batted to follow practically a straight-line path—has a speed of, say, 100 feet per second and is fielded by an outfielder 350 feet from the batter. How many seconds does the outfielder have to get into position to catch the ball? A man on third base can run the 90 feet to home plate at 40 feet per second. A pitcher 65 feet from home can throw the ball to the catcher at a speed of 20 feet per second. Will the runner beat the ball to home plate? How many feet per second does a runner travel if he can cover a mile in 4 minutes?

The distinction between mass and weight can be made and utilized in elementary school work. In daily life we confuse mass and weight and the confusion does no harm. But in these days of space exploration the distinction is important, and it is easily taught. Mass, roughly speaking, is quantity of matter. Weight is the pull that the earth or some other body exerts on mass. Thus, an astronaut on the moon has the same mass as on earth but weighs far less, about one-sixth as much as he does on earth.

Do we have to take children to the moon to demonstrate the difference between mass and weight? Film strips of astronauts walking on the moon would serve. There is also recourse to the child's common experience. An object placed in water is buoyed up by a force equal to the weight of the displaced water. This fact is known as Archimedes' principle. Thus, a swimmer is buoyed up by the weight of the water he displaces, which is just about his weight, and so he is weightless in the water. If a spring whose upper end is held fixed is placed so that the bottom just reaches and is attached to the swimmer, his mass will not extend the spring.

If an object is heavier (in air) than the weight of the water it displaces, it will sink. Water weighs 62.5 lbs. per cubic foot. Given the weight in air and the volume of a piece of lead or other substances, one can readily frame arithmetic problems that ask whether the objects will float or sink in water.

Archimedes' principle applies also to objects in the air and accounts for objects rising instead of sinking. Helium is lighter than air. If a balloon is filled with helium, its weight is less than the weight of the air it displaces. Hence, it is forced upward by the weight of this displaced volume of air, and since its own weight, which pulls it downward, is less, the balloon rises. It is easy here, too, to raise simple arithmetical problems on the motion of the balloon.

A good case can be made for including in elementary school the elements of statistics and probability. The citizen of the United States is faced with statistics in practically every broadcast and in the daily newspapers. The most important decisions of life, as Laplace pointed out almost two hundred years ago, are made on the basis of probability. Even trivial ones, such as the decision to cross the street, are made on the same basis. The interpretation of statistics is by no means simple and the significance of a probability of, say, .75 depends very much on whether it applies to a horse race or a medical treatment. Of course, political leaders and others with axes to grind use statistics and probability to mislead people. Hence, some knowledge of these subjects and their applications to daily life is important to all citizens.

Commercial problems are relevant. Certainly, compound interest should be understood by every citizen, as should the subject of discount. Young children may not be excited by such topics, but perhaps challenging questions would help. If a storekeeper reduces the price of an item by 10 percent and then increases the price by 10 percent, has he restored the original price? Which is the better buy: an item subject first to a discount of 20 percent and then to a discount of 10 percent of the reduced price or the same item subject to a discount of 10 percent and then to a discount of 20 percent of the reduced price?

The elements of geometry are now introduced in elementary school. This is useful knowledge. But as currently taught it seems to be innovation for the sake of innovation. Children are asked to learn any number of terms—acute angle, right angle, obtuse angle, straight angle, and dozens of others—and to know what they mean pictorially. But nothing is done with this knowledge to give it purpose in the eyes of the children. Fortunately, there are dozens of interesting and simple applications.

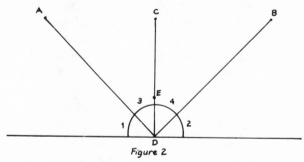

Figure 2

It would be helpful to see some uses of angles beyond their being parts of triangles. One of the commonest applications is to mirrors. A mirror reflects light in accordance with a very simple principle, called the law of reflection. If a ray of light (Figure 2) leaves A and strikes the mirror at D, it will be reflected so that $\angle 1 = \angle 2$. If one introduces the perpendicular CD to the mirror, then one can say that $\angle 3 = \angle 4$. $\angle 3$ is called the angle of incidence and $\angle 4$, the angle of reflection. Hence, the law of reflection says that the angle of incidence equals the angle of reflection. Many rays of light (strictly, an infinity of rays) leave A and strike the mirror at various points. However, only one will reach B—namely, that one for which $\angle 1 = \angle 2$. Hence, a person standing at B sees A in the mirror. In fact, even if CE were a wall between A and B, the person at B will still see A in the mirror.

This law can be readily demonstrated in a classroom by darkening the room, laying a small mirror flat on the desk, and using a pencil flashlight to show the incident and reflected rays. Since every child uses a mirror, the law should be of interest.

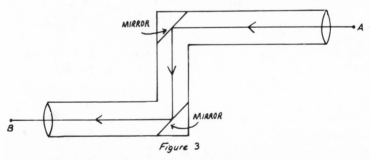

Figure 3

To further impress the student with the importance of angles one can describe or even demonstrate the use of the law of reflection in a periscope. A ray emanating from A (Figure 3) is reflected twice and reaches B.

The law of reflection also applies to billiard balls. If a ball at *A* (Figure 4) is to hit the ball at *B* by being bounced off the side of the table, the ball must be directed so that ∡1 = ∡2.

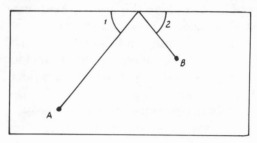

Figure 4

Elementary geometry can be applied to astronomy, preferably with the aid of a model of the solar system, which is commercially available. In this model the sun, the planets, and their moons revolve in precisely the orbits they follow in space and with the correct relative speeds. Such a model is worth far more than any verbal description, diagrams on the blackboard, or pictures in books. A number of mathematical themes—the velocities of the planets, the varying length of the day throughout the year, and eclipses—can be treated readily with the aid of this model. One might object that the subject belongs to astronomy. It belongs to knowledge, and mathematics is essentially involved.

Still another area to which the elements of geometry can be applied is geography. Latitude and longitude are angles formed at the center of the earth. Students can compute distances traveled along a meridian under a change of, say, 5 degrees of latitude or distances traveled along the equator or a circle of latitude under a change of 5 degrees of longitude. (In some instances information such as the radius of a circle of latitude would have to be supplied.)

Some applications are mundane but practical. Students might be asked to consider the areas of various rectangles with the same perimeter. Thus, if the perimeter were 100 feet, the dimensions could be 1 by 49, 2 by 48, 4 by 46, 10 by 40, etc. By using arithmetic and the formula for the area of a rectangle, students would soon discover that the maximum area is given by a square with dimensions 25 by 25. This result is useful. If a farmer wishes to use only 100 feet of fencing, he can have more area for planting if he chooses a square. Or consider a fifty-story office building. The cost is determined mainly by the walls, but the income is determined mainly by the number of square feet of space that can be rented. Thus, the square shape pays off on each floor. Of course other factors, such as the shape of the land, may not allow the use of the square shape, but the closer one comes to it the more floor space will be obtained. Later, on the high school level, the proof that the square furnishes maximum area can be taken up.

Some applications of geometry are not realistic but are surprising. Students are taught (without proof) that the circumference of a circle is 2π times the radius. This formula applied to a circular garden with a radius of 5 feet yields a circumference of $2\pi \cdot 5$ or 10π feet. Suppose one wishes to have a walk 1 foot wide around the garden and a fence to enclose the garden and walk. Then the radius of the circular fence is 6 feet and the circumference of the fence is $2\pi \cdot 6$ or 12π feet. Thus, the fence is 2π feet larger than the circumference of the garden itself. This result is not surprising. But now suppose one were to build a roadway around the equator and the roadway is to be 1 foot higher at each point than the surface of the earth. How much longer is the roadway than the circumference of the earth? Student guesses based solely on intuition will usually be far wide of the mark. The radius of the circle that the roadway forms is the radius r of the earth plus 1 or $(r+1)$. Hence, its circumference is

$C = 2\pi(r+1) = 2\pi r + 2\pi.$

Thus, the roadway is only 2π feet longer than the circumference of the earth—exactly as much longer as the circumference of the fence is longer than the circumference of the garden.

The introduction of scientific applications of mathematics is a lesser problem to the elementary school teacher than to the secondary school teacher. The elementary schools now teach some science, and since the same teacher teaches all the subjects, he or she knows the science involved. Why not teach what is relevant to mathematics in connection with mathematics? The rigid segregation of subjects, which calls for teaching arithmetic at ten o'clock and geography at eleven, is artificial and works against the major goal of education: the integration of all knowledge.

The above examples of real situations—which provide motivation and application as well as the context in which elementary mathematics can be taught—may convince children that the subject, properly approached, is fascinating. As long as the child can think in physical or sensory terms he is at home. Symbols and words acquire meaning only in terms of sense perceptions. Fortunately, teachers and texts at the level of elementary education are becoming committed to the principle that understanding is intuitive rather than logical and are employing intuitive aids such as children's experiences out of school, activities in the classroom, pictures, measurement, and geometrical schemes to answer the question of why we do things the way we do. Conviction derived from seeing that the arithmetical processes do yield what is physically true, rather than proof in the mathematical sense, should be and is becoming the basis for children's acceptance of the processes.

Laboratory materials are now used extensively as aids to pedagogy. The teacher no longer needs to bring apples to school. In the first two or three grades a teacher now has available any number of devices, including Cuisenaire rods, cards, discs, balances, scales, fraction bars, geoboards, clocks, tapes, various types of rulers to perform measurements, geometric models, spinners (to teach probability), and blocks of various sizes, colors, and shapes to represent units, tens, and so on.

Games and puzzles motivate at the lowest elementary school level and it is pleasant to note that these, too, are currently being introduced. For very young children Plato gave this advice over two thousand years ago in *The Republic*: "Do not, then, my friend, keep children to their studies by compulsion but by play." Many commercial organizations are now manufacturing games, and school systems are buying them.

Sadly needed are laboratory materials for the higher grades. One of these, the hand calculator, is coming to the fore and will undoubtedly aid in the learning of arithmetic. (Its precise value will be discussed later.) Laboratory materials, introduced only in very recent years, were advocated by great educators centuries ago. Montaigne and Rousseau advised us to teach the very young with the aid of real things—let children learn by doing.

Through building up elementary mathematics on the basis of physical problems, real phenomena, and laboratory materials, we seek, of course, to teach understanding of the skills in the hope that the understanding will help in remembering the skills and aid in determining where to use them. However, full understanding of our present-day arithmetic is not easily attained, even with the best of pedagogy, because our system of arithmetic is sophisticated—at least for youngsters. Yet the skills should be acquired early in life. The resolution of this difficulty is not to abandon the teaching of understanding but to emphasize the acquisition of the skills. Though the child's understanding may not be complete, or even though he may forget the justifications of the operations, much has been gained by the presentation. Receiving it at least once has a psychological value. It sets the mind at ease, whereas in the absence of understanding the mind remains perplexed and balks at performing. The situation is somewhat the same as when an adult is called upon to donate money to charity. He wants to know why this charity and not some other. Having satisfied himself, he gives freely rather than grudgingly; and in future years he gives without questioning because he remembers not why he chose this particular charity but merely that he had satisfied himself that it was worthwhile. So it is with skills. Having understood why and having learned how to perform them, in the future one performs them without hesitation. One need not and, in fact, should not rethink through the why. As Alfred North Whitehead points out:

> It is a profoundly erroneous truism, repeated by all copybooks, and by eminent people when they are making speeches, that we should cultivate the habit of thinking of what we are doing. The precise opposite is the case. Civilization advances by extending the number of important operations which we can perform without thinking about them. Operations of thought are like cavalry charges in a battle—they are strictly limited in number, they require fresh horses and must only be made at decisive moments.

There are professors who still believe that we can give priority to the concepts and thereby impart understanding. Then, supposedly without much drill, the acquisition of technique will be achieved. Today some go further and argue that the teaching of arithmetic skills can be dispensed with entirely. Students and adults will use hand calculators. As is often the case with new devices, their value is exaggerated beyond all reason. They become the new fad and the excuse for countless worthless papers and talks.

The prospect of hand calculators replacing arithmetic skill conjures up rather ridiculous scenes. Apparently men and women will carry calculators to shop in stores, to decide how much of a tip to leave to waiters and to cab drivers, and to decide how many nickels, dimes, and quarters one needs to pay a subway fare. Perhaps if we were training people only to operate modern cash registers in our grocery stores, the reliance upon calculators would do.

Peculiarly, many of the very same people who favor a curriculum that will prepare students for college mathematics are the ones who would dispense with teaching skills and would substitute the calculator. But students who will go to college must go to high school. And there they will be asked to perform calculations such as $1/a + 1/b$, $a^5 \cdot a^3$, and $(a+b)^2$. The best rationale one can give to students of algebra is that letters stand for numbers, and if one knows how to perform the corresponding operations for numbers, the operations with letters are readily justified and performed. The addition $1/a + 1/b$ is carried out in precisely the same manner as $1/2 + 1/3$. Hence, one must learn the skills of arithmetic to do algebra.

Moreover, even after pushing buttons to perform an arithmetic operation, one should check to see that the answer is reasonable. One might have pushed the wrong button. But estimation as to the reasonableness of an answer calls for knowing a good many arithmetic skills.

The hand calculator can be compared with the typewriter. One can use the latter and dispense with handwriting, but the typewriter does not tell us what to say and how to say it. Similarly, the calculator will not tell us what operations to perform or whether the answer is reasonable. Its value is limited. To one who knows arithmetic and how to use it, the calculator can serve to speed up long series of calculations or calculations with large numbers; it is useful as a check on calculations performed with pencil and paper and thereby serves as a teaching aid; and it does provide a gadget that youngsters can finger and so introduces some novelty. But it cannot replace the teaching of arithmetic skills.

The training of prospective elementary school teachers should, of course, include an understanding of the physical phenomena that may be used to motivate students and to apply arithmetic and geometry, and it should include the use of resources such as laboratory materials. College courses for prospective teachers should also foster many more pedagogical aids, some of which can merely be mentioned here.

Terminology should be and some of it, fortunately, is being improved. In earlier times, when students were taught how to perform the subtraction problem 24-19 they were told to *borrow* 1 from the 2 and to place it alongside the 4 to make 14. They could then proceed. The term *borrow* was a poor one. One might as well have chosen beg or steal. And borrowing, as Polonius said to Laertes, "dulls the edge of husbandry." Moreover, if one borrows, then one should repay. But there is no repayment. The better term is *exchange*. (*Regrouping* is also used.) One takes one of the two tens, converts or exchanges it for ten ones and adds these ten ones to the four. One has exchanged one ten for ten ones, just as one exchanges a dime for ten pennies.

False application of techniques can be a challenging call for critical thinking. Students are taught, rightly, that division of one whole number by another is repeated subtraction. Hence, 7 divided into 28 might be carried out as follows: One tries 1 as a trial quotient and subtracts. This gives

$$
\begin{array}{r}
1 \\
7\overline{)28} \\
7 \\
\hline
21
\end{array}
$$

But now 3 can be the next trial quotient and, in fact, 7 divides evenly three times into 21. The false procedure then is to write

$$
\begin{array}{r}
13 \\
7\overline{)28}
\end{array}
$$

and the answer is presumably 13. Though the flaw may seem trivially obvious to an adult, it is not so to the child. Of course, there are many subtler false arguments and processes that can be employed to advantage.

At some stage in the primary school curriculum students should become fully aware that numbers are abstractions. Simple situations can be used to make the point. A man goes into a shoe store and buys three pairs of shoes at $20 a pair. The salesman says that 3 pairs of shoes at $20 a pair cost $60, and he expects the customer to hand him $60. But the customer instead replies that 3 pairs of shoes at $20 a pair is not $60, but 60 pairs of shoes, and he asks the salesman for the 60 pairs. Is the customer right? As right as the salesman. If pairs of shoes times dollars can yield dollars, then why cannot the same product yield pairs of shoes? (The physicist would give a dimensional argument, but this is a crutch.) The answer is, of course, that we do not multiply shoes by dollars. We abstract the numbers 3 and 20 from the physical situation, multiply to obtain 60, and then interpret the result to suit the physical situation. In this example the clerk's physical interpretation is the correct one, at least in our economy.

The injunction to teach discovery has by now become hackneyed. The fact that it is difficult to do undoubtedly explains why it is rarely done. One must prepare a series of leading questions for each topic and, often without seeming to do so, reformulate a student's reply so that it leads to a useful suggestion. The process is also time-consuming, and many teachers complain that they can't cover all the ground of the course if they teach discovery. But covering ground that succeeds only in causing dislike and even failure certainly accomplishes little. It is not possible here to present the discovery approach to the many topics that have to be taught in the elementary school. But a simple numerical example may serve to illustrate the idea. Students can be presented with the following facts:

$$1 = 1^2$$
$$1 + 3 = 2^2$$
$$1 + 3 + 5 = 3^2$$

They can then be asked to generalize. They need not write the generalization in symbolic form, but they might be able to state in effect that the sum of the first n odd numbers is the square of n.

This conclusion can be supported by interesting diagrams that, in fact, were first used by the Pythagoreans of the sixth century B.C. (Figure 5).

Figure 5

One sees in these diagrams how the addition of an odd number of dots, those above and to the right of the solid lines, produces the next square number. Further, granted the conclusion, the student can be asked to state, without adding, what the sum of

$$1 + 3 + 5 + \ldots + 97 + 99$$

is. There are many such examples, involving only simple properties of whole numbers, that can reasonably call for discovery. The application described earlier of finding the rectangle of maximum area with given perimeter can be posed as a problem calling for discovery of the fact that of all rectangles with the same perimeter, the square has the most area.

Though education in arithmetic has made some progress since the 1970s began, there are still dragons roaming the land, and these must be slain if further progress is to be made. College professors attach very little importance to the courses addressed to prospective elementary school teachers. But these courses are as important as any that professors teach. How these elementary school teachers perform later is crucial. Children start to learn mathematics from the moment they enter school. Their success or lack of success will affect their attitude toward the subject, their confidence in themselves, and, very likely, their attitude toward all learning. Poor mathematics education may be the rotten apple in a barrel of sound ones and by contamination spoil the rest. It is in the course of their elementary school work that far too many people decide that they do not have mathematical minds. Actually, they are victims of teachers who have themselves been victimized by college professors. People "do not have mathematical minds" because they do not receive decent mathematical teaching. Meaningless mathematics will not penetrate any minds. And one who is frustrated by arithmetic will retain an inferiority complex toward all of mathematics, because the subject is cumulative. Even where arithmetic is not a prerequisite, as in large areas of geometry, the mere fact that these are also part of mathematics causes students who have been burned by arithmetic to refuse to let their minds be kindled by new themes.

Mathematicians do like to foster the impression that there is such a thing as a mathematical mind that *eo ipso* is necessarily superior. Though the existence of such minds is highly unlikely, there is the probability that creativity of the Einsteinian caliber does require exceptional abilities. But it is certain that one does not have to have a mathematical mind to *understand* mathematics.

The courses that colleges offer to prospective elementary school teachers are usually a waste of time. Typically, they teach the very same topics that the current so-called liberal arts course offers—the logical development of the real number system; set theory; transfinite numbers; Boolean algebra; truth tables; abstract mathematical structures such as groups, rings, and fields; finite geometries; and a heavy emphasis on axiomatics and proof. In fact, the very same texts that are written for the freshman liberal arts course are advertised and used for the courses addressed to prospective elementary school teachers. Even in the heyday of the New

Math these topics were inappropriate. Elementary school teachers do not need to know this material and even get wrong ideas from it concerning what mathematics is about and how mathematics should be taught.

Despite the shattering effects of past and current education of elementary school teachers, there is reason to be optimistic. Dragons, especially the mythical ones, have been slain in the past. The current educational dragons, even though devoutly believed in by professors, are also myths; these, too, can be slain.

10

Follies of the Marketplace: A Tirade on Texts

Four species of idols beset the human mind: to which (for distinction's sake) we have asssigned names...the third, idols of the market...

Francis Bacon

Curriculum and teachers are the most important factors in education. But there are also texts from which students might learn and which, at the very least, can reinforce the teachers' contribution. Unfortunately, concern for exposition is not one of the hallowed traditions of the mathematical world and the quality of texts at all levels is very low.

The blame for this state of affairs must be laid on the professors. College texts are, of course, written solely by professors. The secondary and elementary school texts are often supposedly cooperative efforts between knowledgeable professors and experienced teachers, but the professors, "obviously" the authorities, dominate the projects. What should we expect of professors insofar as texts are concerned? Many professors are indifferent to pedagogy and others are totally ignorant of it. They receive no training in writing—even of research papers, let alone texts. On the basis of their backgrounds and major concerns one should no more expect effective writing from mathematics professors than good mathematics in mathematical research papers were they written by English professors.

Our expectations are more than fulfilled. Explanations of mathematical steps are usually inadequate—in fact, enigmatic. Because mathematicians do not take the trouble to find out what students should know at any particular level, they do not know how much explanation is called for. But the decision is readily made. It is easier to say less. This decision is reinforced by the mathematician's preference for sparse writing. If challenged, he replies, "Are the facts there?" This is all one should ask. Correctness is the only criterion and any request for more explanation is met by a supercilious stare. Surely one must be stupid to require more explanation. Though brevity proves to be the soul of obscurity, it seems that the one precept about writing that mathematicians take seriously is that brevity is preferable above everything, even comprehensibility. The professor may understand what he writes but to the student he seems to be saying, "I have learned this material and now I defy you to learn it."

Some of the great masters of mathematics did write enigmatically. The most notorious in this respect was Pierre-Simon Laplace. His assistant Jean-Baptiste Biot, who helped Laplace to prepare for the press the latter's masterpiece, the *Mécanique céleste*, reported that Laplace was frequently unable to reproduce the steps by which he had reached a conclusion and so inserted in the manuscript, "It is easy to see that..." Evidently modern textbook writers have taken seriously the precept that one should emulate the masters. Even if one does not become a master thereby, one can at least appear to be one.

There are textbook writers who believe that a mathematical presentation that is logically sound explains itself to the reader who faithfully follows the author step by step. Presumably the meaning need not be stated by the author explicitly but can be grasped by the reader from the details he ploughs through. The authors do not see the need to take the readers into their confidence, to explain where the road is going, why this one is better than another, and what is really achieved. They give no inkling of how a proof was arrived at, why anyone sought the result to begin with, or why anyone should want it now. In effect, the texts are challenges to clairvoyance.

Some textbook writers, unwilling or unprepared to do research, display their "talent" in their texts. They deliberately omit steps that they could not have supplied as students and that they know belong. By pretending that the omitted steps are readily supplied, they seek to put themselves in the position of great masters who have omitted only the trivial. If this condemnation appears too strong, let us remember that with respect to character, mathematicians, whether researchers or teachers, are just a cross section of humanity and, with respect to egotism, a rather disagreeable portion of humanity. In any case, their texts are too often unintelligible.

Surprisingly, many professors object to the few texts that give full explanations and discuss the significance of the ideas being presented. They often complain that such texts are too wordy. Too wordy for whom? These professors prefer to see an enigmatic presentation that leaves the student baffled. Then they, the teachers, can display their brilliance by explaining the text. This preference is well known, but one can also find evidence for it in print. A professor at a good college had the following to say in his review of a text: "So much is here written that is normally spoken by the teacher that the teacher in using the book as a text may find it hard to break away from the book not only in his formal presentation but also in his asides." But if the professor were really teaching ideas and creative thinking he would have so much to do in raising questions, guiding the students' thinking, and improving their suggestions that no book, however helpful, could replace him.

Much poor mathematical writing is due to sheer laziness. There are mathematicians who fail to clarify their own thinking and attempt to conceal their vagueness by such remarks as, "It is obvious that...," "Clearly it follows that...," and the like. If a conclusion is really evident, it is rarely necessary to say so; and when most authors do say so, it is surely not evident. Often what is asserted as obvious is not quite correct, and the unfortunate reader is obliged to spend endless time trying to establish what does truly follow.

In some cases the difficulty is the writers' sheer ignorance. Even matters that are well understood by reasonably good mathematicians are not understood by numerous authors of high school and college texts. They put out books mainly by assembling passages and chapters from other books, and where the sources are inadequate so is the pirated material. In their books one finds inaccurate statements of theorems, assertions that are not at all true, incomplete proofs, failure to consider all cases of a proof, the use of concepts that are not defined, reliance upon prior results that were not proved or are proved only subsequently, the use of hypotheses that are not stated, two nonequivalent definitions of the same concept, extraneous definitions, assertions of a theorem and its converse with proof of only one part, and actual errors of logical reasoning.

A glaring deficiency of mathematics texts is the absence of motivation. The authors plunge into their subjects as though pursued by hungry lions. A typical introduction to a book or a chapter might read, "We shall now study linear vector spaces. A linear vector space is one which satisfies the following conditions . . ." The conditions are then stated and are followed almost immediately by theorems. Why anyone should study linear vector spaces and where the conditions come from are not discussed. The student, hurled into this strange space, is lost and cannot find his way.

Some introductions are not quite so abrupt. One finds the enlightening statement, "It might be well at this point to discuss . . ." Perhaps it is well enough for the author, but the student doesn't usually feel well about the ensuing discussion. A common variation of this opening states, "It is natural to ask . . . ," and this is followed by a question that even the most curious person would not think to ask.

One need not always precede the treatment of a mathematical theme with the major reason for studying it. The introduction could be the historical reason the topic was studied. But if this is not the reason for its importance today, applications of current importance should immediately follow the treatment. Unfortunately for the authors, most applications involve physical science, with which few mathematicians and high school teachers are familiar. The best that many authors can do is simply to state that there are important practical applications of the subject treated. Some actually promise they will discuss applications but fail to do so.

Problems of science need not be the sole motivation. The mathematics to be taught might be related to the students' world. Perseverance would reveal what excites student interest, but this effort is far more than authors are willing to undertake. The consequence is bare bones and no meat. Even the prettiest woman seeks to enhance her appearance with dress and cosmetics. Similarly, mathematics should be made more attractive by relating it to the interests of neophytes.

The lack of motivation has been criticized by Richard

Courant, whom we have cited in other connections:

> It has always been a temptation for mathematicians to
> present the crystallized product of their thoughts as a
> deductive general theory and to relegate the individual
> mathematical phenomenon to the role of an example.
> The reader who submits to the dogmatic form will be
> easily indoctrinated. Enlightenment, however, must
> come from an understanding of motives....

To begin a text with a statement of the axioms is to write a
work that omits the first chapter and that demands of the
reader an understanding without which he cannot comprehend the text before his eyes.

Mathematicians wreak additional hardships on the
students by indulging their own tastes. As mathematicians
they recognize the advantages of generality and abstraction.
Surely a general result covers many special cases. Hence,
they conclude, it is more efficient to present the general
result at once. Logically this is correct; pedagogically it is
false. The recklessness with which authors of texts plunge
into generalities indicts their judgment.

For example, before students have worked with concrete
functions such as $y = 2x$, $y = 2x+3$, $y = x^2$ and the like, they are
asked to learn a general definition of function in terms of
mappings. A mapping from a set A to a set B is a set of
ordered pairs with each first component from set A and each
second component from set B. Mappings are "wonderfully"
broad. The relationship of a set of fathers to a set of sons is a
mapping, and knowledge of this fact "clearly" improves the
relationship between parents and children. To be sure, the
definition of a mapping includes the concrete functions just
mentioned; but it also includes relationships that students
will never encounter or that are certainly not illuminated by
the mathematical definition. Moreover, the vagueness of a

general definition leaves students uneasy.

Whereas a generalization extends to a wider class of objects a result known only for a special class—for example, one may prove for all triangles a theorem known previously only for isosceles triangles—abstraction selects from different classes of objects properties common to the classes and studies the implication of these properties. No one would question the value of abstractions for mathematics, but one must question the abstract approach to the concrete. Children do not learn about dogs by starting with a study of quadrupeds. This elementary principle does not seem to have been learned by mathematicians. They love abstractions and indulge in them freely, of course at the expense of the student.

Many other faults already cited apropos of teaching methods are repeated in the texts. Rigorous presentations addressed to beginners in a particular subject are common. From a conceptual standpoint the most difficult mathematical subject is calculus. The concepts can be far more readily understood intuitively, and this is how mathematicians grasped the subject until, after two hundred years of effort, they managed to erect the proper logical foundations (Chapter 7). But many modern authors courageously risk the students' necks. They start their calculus texts with the rigorous formulations of the concepts and at the outset succeed in destroying the students' confidence in their ability to master the subject.

Some authors choose the rigorous approach because they are insecure. They are fearful that if they compromise in order to help the student they will appear ignorant to their colleagues. To justify their stand they argue that texts at least must be precise and complete, and often they insist that understanding can best be obtained through the rigorous formulation. The result is elementary mathematics from a complicated standpoint.

Most authors profited as little from their study of English as their readers then profit from the study of mathematics. The writing in mathematics texts is not only laconic to a fault; it is cold, monotonous, dry, dull, and even ungrammatical. The author seeks to remain impersonal and objective. As one reviewer said of the writing in a particular text. "The book is mathematically masterful, grammatically grim, literarily limp, and pedantically pompous. It tells the undergraduate more than he wants to know, presuming, at the same time, that he knows more than he does." Of course good texts should have a lively style, arouse interest, and keep the readers' background in mind. But few do. The books are not only printed by machines; they are written by machines.

One ingredient of style is humor. A relevant story or joke does revive a sagging spirit. But the professors object. They use the same text a number of times and to them the humor becomes stale or the joke palls. But for whom is the book written? What would these professors say about an actor who must repeat for the thousandth time a most dramatic line or a joke as though it were the first time he ever said it?

Beyond their sheer incapacity or unwillingness to write interesting mathematics, authors splurge in terminology that baffles the reader. In addition to using many technical words unnecessarily, mathematicians love to introduce new vocabulary. Thus they have long used the words *homomorphism* and *isomorphism*, which at least preserve the etymological meanings of similar structure and same structure, respectively. Instead of saying that a certain homomorphism is an isomorphism, the practice is to say that the homomorphism is faithful—a statement that does nothing to convey its meaning in mathematics, though it may have the merit of suggesting a steadfast, if illicit, romance. Of course, the appearance of a new term gives the impression that a new concept has been introduced. The terms *greatest lower bound* and *least upper bound* of, say, a set of numbers were used for years and do describe what they stand for; now, presumably in the interest of brevity and certainly in the direction of making comprehension more difficult, the terms *infimum* and *supremum* are used. *Single-valued functions* and *multiple-valued functions* are now *functions* and *relations*. Further, one no longer speaks of the values of x that satisfy $x^2 = x + 7$ but of "truth values." Apparently, truths can now be obtained readily and we need no longer ponder the mysteries of the universe. New terms to replace old ones appear constantly. This practice disturbed even Cauchy, who shared with Gauss leadership in mathematics in the first half of the nineteenth century, and he felt obliged to complain of the strange terminology introduced in his day: "One should enable science to make a great advance if one is to burden it with many new terms and to require that readers follow you in research that offers so much that is strange."

Instead of introducing new, meaningless terminology in place of suitable words, mathematicians would do well to replace the old terminology that has misled students. Terms such as irrational, negative, imaginary, and complex, which were historically terms of rejection, remain in the lexicon of mathematics to disturb students. Even the great mathematicians of the past were frustrated by such terms. The resistance to imaginary numbers persisted for three hundred years after their introduction, partly because the word *imaginary* suggested something unacceptable. As Gauss remarked, if the units 1, -1, and $\sqrt{-1}$ had not been given the names positive, negative, and imaginary units but were called direct, inverse, and lateral units, people would not have gotten the impression that there was some dark mystery in them.

Meaningless terminology is only one evil of the language used. Mathematicians believe in brevity so much that they invent shortened terms. A partially ordered set is now a *poset*. Still more brevity is achieved by using acronyms a.e. (almost everywhere). This atrocity, added to already barbarically poor writing, makes it almost impossible to read the text, let alone understand it. The l.q.m.w. (low quality of mathematical writing) has no bottom.

The evils of terminology are compounded by the excessive use of symbols. No one would deny that mathematical thinking and processes are expedited by the use of symbols, and one of the great features in the progress of mathematics was the introduction of better and better symbolism. But mathematicians have turned a virtue into a vice. They sprawl symbols over all the pages of their texts just as some modern painters splash paint on canvas. Page after page, almost devoid of exposition, is filled with Greek, German, and English letters, and various other symbols. Some books use as many as a few hundred symbols, presumably in the interest of brevity but more likely to conceal shallowness. Even a gain in brevity hardly compensates for the burden on the reader's memory. What is worse is that a symbol introduced on page 50 is not used again until page 350, with no reminder to the reader of what the symbol stands for. A few authors, somewhat conscious of this problem, include in their texts a glossary of the symbols employed. However, the reader stuck on page 350 must interrupt his reading to find the meaning of the symbol among the several hundred in the list. The natural reaction is exasperation. In modern texts symbols do not facilitate communication; they hinder it.

Many authors seem to believe that symbols express ideas that words cannot. But the symbolism is invented by human beings to express their thoughts. The symbols cannot transcend the thoughts. Hence, the thoughts should first be stated and then the symbolic version might be introduced where symbols are really expeditious. Instead, one finds masses of symbols and little verbal expression of the underlying thought.

As in the case of terminology, much could be done to improve older symbolism. Perhaps the most imperative need is to replace the symbol dy/dx of calculus. The most important idea to be transmitted in calculus is that the derivative is not a quotient but the limit of a quotient. However, the symbol dy/dx, though intended to be taken as a whole, looks like a quotient of dy by dx. (In fact, this is what it was for Leibniz, who failed to formulate the precise concept.) Hence, the notation seems to refute what the teacher must attempt to convey. Superior notations have been proposed, but professors resist change in this area as zealously and perversely as they promote it where the traditional symbolism is altogether adequate.

A mathematician of the sixteenth or seventeenth century often presented his discoveries in the form of an anagram that was intended as evidence to his rivals that he had solved a problem but that was also intended to be undecipherable to the rivals so that they could not claim they, too, had solved the problem. When challenged, the composer of the anagram could then reveal what the anagram stood for and establish his claim. This practice continues today, except that the anagram is called a textbook.

Mathematicians claim to teach thinking, and this can be promoted by getting students to help discover theorems and proofs. But the texts do no such thing. Definitions, axioms, theorems, proofs, and obscurity are the style and content, the sum and substance. This type of presentation has the advantage—for the author—of facilitating the writing. One does not have to think about what to say because the theorems and proofs of the usual undergraduate textbook are well known. However, as we have previously noted (Chapter 6), most theorems of any consequence have been reproven many times; each time some refinement or modification is achieved that makes the theorem more general or the proof shorter. Often an ingenious trick will do the latter. Since even the original proof may have been the product of weeks, months, and perhaps years of thought, to which a succession of mathematicians may have contributed, a modern proof is almost sure to be sophisticated and highly artificial, though mathematicians would describe it as elegant. These refined proofs, presented in a page or so, stun and humble the students. They cannot help imagining themselves being called upon to make such proofs and readily realize that the task would be inordinate for them. The inevitable consequence is that they lose confidence in their ability. To pass examinations they memorize the proofs.

Good writing, like good teaching, calls for letting the students in on the struggles mathematicians have undergone to arrive at the proofs. Students should be told how long and hard the best mathematicians worked to obtain the proofs, and how many false proofs were often published in the belief that they were correct. This history not only would avert discouragement and loss of confidence but also would dispose students to the kind of effort they must be prepared to make when attempting a proof on their own.* But authors are reluctant to level with the students. By presenting proof after proof with no mention of how these were obtained, the authors seem to suggest that the clever proofs are due to them and, very likely, this is the impression some wish to give. Authors do not recognize the psychological damage of a bare logical presentation.

*   Texts do present historical material, usually to the effect that Descartes was born in 1596, died in 1650, and had one illegitimate child.

Why can't texts be more informal, almost conversational? Suppose an author is about to present the theorem that the three altitudes of a triangle go through the same point. This fact should surprise anyone. Should not the author remark on this, perhaps state that it is not an expected fact, and first give some intuitive reason that it should be so before proving it? Calculus texts treat the derivative of the product of two functions. Students expect that the result should be the product of the derivatives and are surprised to find that it is not so. Even Leibniz stumbled on this point and spent a whole month getting the correct result. The authors could state what superficial argument suggests and then point out why it is not correct. The fact that Leibniz struggled to understand this matter might also be mentioned and would reassure students that they are not so far below the Leibnizes in intellectual capacity. Such discussions should precede the formal presentation.

To rebut the charge that the texts proper do not call for student participation and thinking, the authors point to the exercises. Good exercises could be some redress for the dogmatic text. However, the texts usually work out half a dozen typical problems in each section and then assign exercises of the same type. The students, called upon to do an exercise, look among the illustrative examples to find one that fits the exercise. They then repeat the steps made in the illustrative example without necessarily understanding them and certainly without having to do any thinking for themselves. Thus, the students do the homework successfully and feel satisfied. The professors, in turn, congratulate themselves on their successful pedagogy.

Of course, some illustrative examples are needed. Students cannot be expected to acquire techniques without guidance. But the examples should be accompanied by a discussion of how the theory is involved, why the solution should take one course rather than another, and any other pertinent comments. In fact, rather than being set out as examples, these illustrations are best incorporated in the text proper to oblige the students to read the text—something that students often shirk if not compelled to do it. Moreover, some of the exercises could raise questions about the examples. Alternative methods might be proposed that may or may not work, and the student might be asked to evaluate them. Mere repetition of a process that is illustrated will teach technique, but it will not inculcate understanding or foster thinking. The usual exercises are intellectual slavery rather than intellectual challenges.

Many textbook authors boast of the number of illustrative examples their texts contain. What they are really saying is that the students do not have to read the texts or do any thinking. These texts are rightly called cookbooks. Actual cookbooks usually offer recipes: pour a half cup of flour into a bowl, add one quarter cup yeast, sprinkle with vinegar and bake for one hour. Lo and behold, a cake appears. But the cookbook gives the cook no idea why such a mixture produces a cake. The illustrative examples are likewise recipes for getting answers. If the recipes were changed and produced absurd answers, the students would not have the insight to recognize this fact and would be content as long as their answers agreed with those given by the text. Clearly, the valuable role that texts could play in the educational world is nullified by the various defects we have cited. The students certainly do not read the texts, because the texts are unreadable.

Surprisingly, when choosing texts for their classes many professors have said openly that they do not care about text exposition. They look only at the illustrative examples and the exercises. But in our civilization, learning to think and learning to use books are surely some of the objectives of higher education. Apparently these objectives have been abandoned.

Since the texts are so bad, one is impelled to ask, why are such texts chosen? The reasons are numerous. The poor exposition is not recognized by most professors because they themselves are not trained in writing. Lack of motivation and application in textbooks is even welcomed by professors. Such material must usually draw on subject matter that lies outside mathematics proper, and to teach it would require that the professors know and feel secure about, for example, a bit of science. But the professors do not know science, and they are not willing to learn it just to do a better teaching job. In fact, many professors fear any book that would make them deal with the history of mathematics, science, or cultural influences. Hence, they choose one that takes the straight and narrow path of mechanical, technical mathematics and routine exercises. Terminology, symbolism, rigor—these are dear to the hearts of mathematicians. From their point of view one could not possibly overdo such features.

A major reason for the choice of poor college texts is that the bulk of the undergraduate courses is taught by graduate students. For such teachers, stock material and routine presentations are musts. Any felicitous or unusual approach, especially if it calls for pedagogical skill, would be ignored or bungled.

Many professors choose a text because the topics treated are what they want to teach. But they do not care about the text's presentation. They give their own. The student is then faced with the task of reconciling what the teacher says and what the book says. This is difficult to do in mathematics. And since the book is most likely to be poor, the difficulty is compounded. If the professor really has a better presentation than what is available in an existing text, he should write up his material and distribute it so that students will not have to spend the class time in copying. In many cases the professor's presentation is not better, but he considers it demeaning to follow someone else's.

There are even university professors who deliberately adopt a difficult book because it bolsters their ego to be able to say that they are using it. They hope that others will judge them and their students favorably, because presumably both can master such a book. Actually, many of the professors who choose such books are hard put to understand them, but the students are so much more bewildered that the professors can get away with almost any kind of explanation.

Professors at four-year colleges often feel inferior to those at the prestigious universities. To overcome the feeling of inferiority many four-year college professors try to "outdo" the university professors by adopting texts that are far too difficult for the students. When asked at a professional society meeting what texts they are using, they can name them and imply that they are really doing wonders with their students and, of course, have no trouble themselves in teaching on the advanced and sophisticated levels that these texts bespeak.

Teachers at the two-year community and junior colleges, which, on the whole, have the weakest students, also use difficult texts just to be able to boast that they are teaching on a high level. They claim they must use these books to prepare students who will transfer to a four-year college. But they kill off the students and so make transfer impossible. Only about 25 percent go on to a four-year college, and most of these students do not take any more mathematics. The phenomenon of low-level institutions using high-level texts is especially prevalent in areas dominated by a major university.

Many texts are chosen by a committee. If a book contains

any applications, some professors who are unfamiliar with them will object. Other professors may rightly or wrongly object to the level of presentation. Still others may object to the "wordiness." The consequence is a compromise that almost necessarily is a dull, meaningless, inept book. Even where a text is chosen for departmental use by a single professor, he may, like so many others, lack insight, conviction, and determination, and pick a "safe" book—which usually means a mediocre one that will satisfy most of the professors.

Quite often a department decides to change the text in use because some members complain that it is not satisfactory; or it may be going out of print. One would think it might be replaced by a good text, but that rarely happens. Most of the staff prefer a text that is old hat, so they do not have to read it and do new exercises. Hence, the replacement is usually a "copy" of the previous text. After all, change is sufficient evidence of progress in our society.

There are many other reasons that a professor will pick a poor text. Professors are most likely to be narrow specialists. An algebraist who is called upon to teach differential equations is not interested in how to teach that subject. He wants a book that is easy to teach from, and this means one that presents either a series of techniques or a canned sequence of theorems and proofs that need only be repeated.

Even if all or a majority of the texts were good, the students in many institutions would still suffer. Some professors choose texts that interest them and from which they can learn new ideas or new proofs, whether or not these texts are right for the students. Thus, an algebraist might pick an advanced text for an elementary course not because the students can learn from it, but because he can. At one respectable institution the professors used a text that was two or three levels higher than the course, and a large percentage of good students failed. Others, also highly qualified, became discouraged and abandoned mathematics. When the professors were asked why they used such an unsuitable text they replied that they were conducting an experiment. They might just as well have said that they had fired six bullets into a man's heart to see if he would die.

Why are so many poor texts written? The main reason is obvious—greed. Texts bring in royalties, and money does interest some people. To make money, one must write a text that sells well. But the poor texts sell best, and the money-minded author caters to the market. Most professors write with more than one eye on the market. They rivet their attention on it. What happens is well illustrated by the history of calculus texts. For years only mechanical or cookbook treatments of calculus were used. Authors, accordingly, wrote cookbooks. As American professors became better educated, they decided that students should receive the benefit of professorial enlightenment and that calculus should be taught with a full background of theory. A spate of rigorous calculus books soon appeared on the market. When this pedagogical blunder became apparent and the intuitive approach became popular, professors showed their open-mindedness and flexibility by turning to an intuitive approach. It did not take long before the very authors of rigorous texts wrote intuitively oriented texts and even boasted that they offered this approach. Professors do learn remarkably fast—what the market wants.

Because most authors aim for the largest possible market, they repeat endlessly books that sell well. All that is required is a minimum of knowledge, shoddy writing, standard exercises, and reasonable caution against outright plagiarism. One need only vary the order of the topics to make a book seem different, and since there are about twenty-five topics in the usual text, the possible permutations are large enough to allow for many thousands of "different" college algebra, trigonometry, calculus, and other texts to be written. The fact that the sources may be incorrect or poorly written is a minor concern compared to the expected gain. To hide obvious repetition of existing texts, some authors introduce a few variations, such as contrived proofs even though more natural ones are available, sophisticated definitions, new terminology, and their own brand of symbolism. When accused of plagiarism the professors can always retort that the truth never changes.

One must of course have a different title. But then one can use *College Algebra, Elementary College Algebra, College Algebra: A Full Course, College Algebra: A Short Course,* and *College Algebra: A Seven-Eighths Course.* The possibilities are clearly infinite. In fact, since there are irrational numbers, one could use *Algebra: An Irrational Course.*

The outright imitation of successful texts—successful financially though usually not at all pedagogically—is a fact. Many authors do not hesitate to admit this. They speak proudly of their books as being in the mainstream of mathematics education, as though this fact is an assurance of quality. Actually, in view of what books sell best, a book in the mainstream is sure to be dull, unoriginal, and pedagogically disastrous.

Are all texts repetitious of each other? No. Another spate of bad texts comes from professors who have achieved a reputation for research in their specialty or whose name is well known in the mathematical world, perhaps because they have held high office in a professional society. These authors, most of whom have never or only rarely taught the courses for which the books are intended and are unconversant with how college students think and what backgrounds the students have, nevertheless decide to cash in on their names and plunge unhesitatingly into the writing of texts. "Genius" transcends mediocrity; so these texts contain innovations in concepts and proofs that students cannot possibly grasp. The exposition of the topics is shoddy and the writing is shameful. The books are hastily written and often contain numerous errors. Chapters or sections begin with one objective and end up with another. Within the same section authors shift from one topic to a totally unrelated one. They ask the students to do exercises that are not workable on the basis of the material in the text or, if related to the text, require a Newton. To make a token gesture to that sector of the market that wants some applications, these researchers include some brief mention of relativity or quantum mechanics, topics that mean nothing to undergraduates at the levels for which the texts are written. It is clear that these professors dash off the books as fast as they can just to get them out and "earn" royalties. Were these authors judged by their texts they would not be admitted as graduate students to any decent graduate school. Nevertheless, many schools adopt such texts on the basis of name alone. Usually the texts are so bad that they are dropped after one year's use. About all one can say of them is that they are flops *d'estime*. Ironically, these prestigious professors, who rush to write texts for low-level courses, would disdain teaching them or, if obliged to do so, would be ashamed to admit that they were teaching such lowly work.

Sometimes these prestigious professors resort to second and third editions and, having learned by this time how to meet the market on its terms, sell more books. The venality of such professors and their crass commercialism are disgraceful. In these later editions they may succeed in selling more books, but they also succeed in sacrificing students and vitiating educational goals. Cheap fiction, potboilers, are far more excusable because the authors make no pretense to ethical principles and are not under any obligation to develop young minds. If these professors are really capable research men or seek to exert beneficial influence through office-holding in professional societies, why do they lower themselves by writing the hundredth facsimile of cheap, commercial texts? Or are the supposedly intelligent professors as badly confused about their role and goals in life as any adolescent?

The problem of writing for financial gain does call for keeping up with the market. As we have already observed (Chapter 7), mathematical teaching as well as mathematical research is swept by fads. Analytic geometry, formerly taught as an independent course preceding calculus, is now submerged in calculus. The successful author must yield to this fad or his book will not sell. If the fad is to incorporate linear algebra in the calculus or the differential equations text, whether or not there is any point to doing so, one must incorporate the linear algebra. To keep up with fads one must put out new editions every few years. But professors do not object because this eliminates the secondhand market for the older edition and students are obliged to buy the new one.

The determination of what the market wants is made rather scientifically. The publishers canvass the colleges for what they would like to see in the texts, and then the authors willingly set about supplying the common denominator of those wants. The author's own convictions, if he has any, as to what a text should contain are irrelevant.

The normal market can be enlarged by special devices. One such device is to offer applications but to crowd them all into the last chapter. There is method in this madness. Applications are desired by some professors but frighten off others. If they are placed at the end, professors who do not want to teach them manage to end the semester before reaching the last chapter, thereby omitting them with least embarrassment. Applications placed at the end of a text serve little purpose in any case, because whatever value these applications might have as motivation and meaning for the mathematics proper comes too late.

To enlarge their market many authors employ ruses that are deliberately fraudulent. When the New Mathematics became popular these authors took traditional books, inserted a few pages of New Mathematics material here and there, changed terminology in spots, and sprinkled words such as *sets*, *commutative law*, *inverse*, and the like throughout; they then proclaimed that they were presenting the New Mathematics. Many teachers aided in this fraud because they could convince their superiors that they were teaching the New Mathematics, while actually continuing to teach the material they either preferred or knew better. Calculus texts often contain a facade of rigor to please those professors who wish to include some theory but the rigor, usually in the first chapter, is thereafter never utilized.

There are other types of deception. One would expect that a text entitled *Mathematics for Biologists* would contain not only the mathematics that biologists use but also some indication of how biologists use it. But the contents are the same as any traditional text that covers the same level of mathematics.

Many authors know that students come to college disliking mathematics. However, some colleges still require a course in mathematics as a degree requirement. Even if they don't, the professors wish to attract students to a mathematics elective so there will be more jobs. Hence, many authors write texts that purportedly offer an appreciation of the role of mathematics in our civilization. The titles are inviting: *Mathematics, An Intellectual*

Endeavor; *Mathematics, the Science of Reasoning*; *An Appreciation of Mathematics*; *Mathematics, the Creative Art*; *Mathematics, Art and Science*. But the texts teach axiomatics, symbolic logic, set theory, topics of the theory of numbers such as congruences, the binary number system, finite geometries, matrices, groups, and fields and so do not really live up to their promise (Chapter 6). Clearly, one can't judge a book by its title.

Since a course in mathematics proper does not attract liberal arts students and has little value for them, some professors have taken another tack. For their books they gather together curiosities, trivia, puzzles, and bits and pieces of standard topics that never get to any serious level and do not require any thinking on the part of the student. The chapters are deliberately unrelated to each other so that the student will not have to carry an extended train of thought and so that the teacher can pick and choose what pleases him. Since these measures rarely succeed in interesting students, some professors have resorted to the ingenious device of including cartoons. There are even calculus texts "enlivened" by cartoons. Why not? After all, isn't mathematics supposed to be fun? Perhaps pointed, truly humorous cartoons can be admitted as a pedagogical device on the college level, but shallow sequences of drawings that would hardly elicit a smile from six-year-olds make no contribution. Something can be said for cartoons: They do enlighten us as to the intellectual level of the authors. Mathematical texts do not as yet resort to pornography, though this means of attracting students would be more acceptable because it would not be mistaken for a pretense to education in mathematics.

These puerile "liberal arts" texts also sell well. Students, deceived or not as to the worth of the material, can earn credit for the course without really being pressed into thinking. The professors can teach such material without any effort and thereby "solve" the problem of what to teach the liberal arts student.

Many professors express concern that the lowering of standards for admission to college and automatic admission of any high school graduate will result in the lowering of the educational level for all students. But by publishing the liberal arts and cookbook texts we have described, which are used in hundreds of colleges, these professors have reduced standards to about the lowest level possible.

The financial gain to be derived from textbook writing has corrupted many professors. Some authors ask publishers for guarantees up to $100,000. Apparently the authors have no confidence in their works and seek to ensure profit by a guarantee. The argument is sometimes made by authors that a publisher will work harder to push a book on which it has given a guarantee. But this is hardly a justification. The investment of the publisher can range from $50,000 to $100,000. Surely no publisher will invest such a sum and then fail to promote the book. If the author is asking for sales promotion beyond the merits of the book, he is certainly culpable.

There are two possible controls over the quality of texts. The first is reviews. Most texts are reviewed in one or more of the professional journals. However, some reviewers are apparently too polite to write the condemnation that most texts should receive. Instead, they merely describe the contents or compare the book with similar ones and often end with the "compliment" that it is good because it is just like the others on the market. Other reviewers respect the principle of honor among malefactors. What we need is honest, damning reviews of books that impose unnecessary hardships on students and fail to teach the values that the course in question should offer.

Though critical reviews of texts are rare, one does find some. One reviewer of a new calculus text said that the "exercises are presented with less imagination than"—and here he mentioned a best-selling calculus book—"if that is possible."

One might look to the publishers to control the quality of texts. But this is not fair. Publishers do have manuscripts reviewed before they accept them; but generally the reviewers are men who teach at the same level as the proposed book, and they are no more critical and no more demanding about pedagogy than the authors. If they see material they are able to teach, they approve the manuscripts.

Moreover, publishers are in business to make money. This is their avowed purpose, and they cannot alter the market. If they do not yield to it, they will fail. They certainly cannot exist on the rewards of virtue. No doubt many exaggerate the qualities of what they publish. Often, too, they seek to anticipate a trend and accelerate it by promoting books that further it and give the impression that such texts are already in wide demand. They did this when the New Mathematics was in the offing and are now hastening to break from the New Math because they foresee its doom. Publishers are often criticized because they publish books just like dozens of others already on the market. But if a publisher is to stay in business he must have saleable books in each of the subjects and perforce must duplicate existing books. The better publishers do compensate somewhat for publishing junk by putting out high-quality monographs and treatises on which they lose money though they may gain prestige.

The responsibility for good texts definitely rests with the professors, who, unfortunately, regard their station as practically a license to publish. In these times the only concession they must make to secure the full imprimatur is to have an opening chapter on set theory, whether or not it is relevant to the body of the book or referred to in later chapters.

Good texts, so sorely needed, would raise the educational level immeasurably. Not only students would benefit. Young teachers, older ones when called upon to teach a course in an area unfamiliar to them, and even knowledgeable and competent teachers can learn much from a good text because the author would have devoted months and years to the selection and presentation of the material, whereas the teacher could not hope to do that in more than one area.

The low quality of the texts is the severest indictment of the professors. Those who deliberately cater to the market even when capable of doing a better job besmirch their character. Those who write texts for courses they have never or only rarely taught impugn their integrity. And experienced teachers who sincerely attempt to write well demonstrate that the arts of pedagogy and writing are rare gifts.

This derogation of the quality of American mathematics texts may seem overdrawn or grossly exaggerated. It is not. The low quality is as much a consequence of the development of the nation's educational efforts as are the poor content and pedagogy of the courses and curricula. The principle of universal education from the elementary school to the highest levels students can attain certainly was and is desirable. But the constant immigration of mainly poor and uneducated people has placed a burden on the country that would be difficult to carry under any conditions. To make matters worse, the emphasis on research in the last thirty or forty years has diverted manpower from teaching and so has cut off the flow into the fountain of all our educational efforts, the teaching in the colleges. Perhaps rather belatedly we shall develop sincere and capable cultivators of mathematics—a science and an art—who will recognize that exposition is as vital in their medium as it is in painting, music, and literature.

11

Some Mandatory Reforms

When I see how much education can be reformed, I have hope that society may be reformed.

Gottfried Wilhelm Leibniz

The defects in our educational system cannot be eliminated by one measure. There is no one cure for all diseases. Yet, little by little, medicine has conquered some and alleviated the gravity of others. In the educational field the universities' insistence on research as the qualification for appointment and tenure of professors (despite the low quality of much of the research and its irrelevance to teaching), large lecture classes, the use of teaching assistants on a wide scale, and inadequate textbooks are all highly detrimental to the progress of mathematics and to the effectiveness of education. Some helpful steps are apparent, and we have to be willing to take them.

The first remedy lies in recognizing scholarship as well as research. Research in mathematics means the creation of new results or, at least, new methods of proof. Scholarship—which fundamentally implies breadth, knowledge in depth, and a critical attitude toward that knowledge—is currently deprecated. This distinction is not made in the social sciences, the arts, and the humanities. The person who digs up facts about an older civilization, who writes a detailed and perhaps critical biography of some major or minor historical or literary figure, or who puts together various theories of economics or government is considered creative, though there may be no single new fact in a given work. Of course, there are seminal thinkers in the nonmathematical fields. Some of their work is as novel, as creative, as anything produced in mathematics. But the distinction between old and new cannot be made as readily. In any case, in these fields new work is only a small part of what is accepted and even honored as research. The re-search of what has been done is accorded as much distinction as new work. In fact, a critical biography or evaluation of a man or an era is often lauded far more than the man or men whose work is being assessed. A lucid explanation or interpretation of mathematical research is worth far more than most research papers. Unfortunately, such presentations, even if of high quality, are held in low esteem. But it is scholars—people with a deep and broad knowledge of mathematics and an ability to communicate whether or not they contribute new results—who can correct many evils and perform many vital tasks.

In recent times the overemphasis on research has forced the production of tens of thousands of papers. Beyond the unreasonableness of compelling teachers and scholars to do research, it has also obliged mathematicians to seek specialties so that they can keep abreast of what is being done in the area in which they seek to publish. The consequences have been a proliferation of obtuse papers of dubious value and the fragmentation of mathematics into an incoherent mass of details. More than ever, it has become necessary to decipher the cryptic research papers, to salvage the gems from the sludge, to connect in a coherent account the mass of disconnected results appearing in the hundreds of journals, and to give prominence to deserving contributions.

More specifically, scholars can elucidate the inscrutable results contained in research papers. Because they can spend more time reading the published papers, they also can detect duplication of results; and the very knowledge that such duplication will be detected would deter those who would consciously publish old results disguised in new terminology and symbolism. Scholars can serve as critics of writing and here, too, force authors and publishers who would fear criticism to aim higher.

There are other valuable functions for the scholar. He can write an expository paper on results obtained in one area and so make the methodology and results accessible to those in other fields. Such a paper can tie together in one intelligible account many research articles that individually mean little or that are of discernible significance only to the specialist. Expository papers would not only broaden the knowledge of currently narrow researchers but might also reveal and synthesize the relationships between works in different areas. They might, in addition, aid scientists and engineers to learn about work that can be useful to them. Survey papers do appear occasionally in the literature, but these are written by specialists for specialists in one particular area and do not aid outsiders.

Today researchers hold innumerable conferences on specialized topics. Physicists and engineers wisely do not attend the usual conference, because it is unintelligible. Instead, they spend time creating what they need, even though their creations may already exist in the mathematical literature. Scholars could lead conferences that explain the research to a broad spectrum of mathematicians, physicists, and engineers.

Lucid, critical, perceptive synthesis, so sorely needed in

the present deluge of publications, would evaluate the import of the findings of the various specialized disciplines and assess them in the light of the overriding questions which prompted the development of the specializations in the first place and which parochial specialists have often lost sight of. Thereby synthesis and synthesizers would keep alive and in the forefront what the discipline as a whole is trying to do or should be doing. When one thinks of the massive manpower, money, and space devoted to mathematical research and the conflict between it and education, then certainly the evaluation of mathematical research and the determination of for what and for whom it is intended is of vast importance.

Scholars, raising questions about the worth or direction of a particular specialization, would keep alive the spirit and activity of dissent. Specialists should be called upon to defend their activity and not be permitted to hide behind such shibboleths as "I am creating art," or "I like what I do." There is a function for the gadfly who poses questions that many specialists would like to overlook. Polemics are healthy. They not only tend to reduce fashions to a scale that their worth warrants, but they also may support the worthy unfashionable and the only seemingly ridiculous. Without scholarship—the organization, explanation, interpretation, and criticism of research—the currently vast number of proliferating disciplines steadily gain in quantity as they lose in quality, vision, and effective use of the little in them that is worthwhile.

Scholars would, by definition, be thoroughly informed in the history of mathematics, which has many lessons that should be brought to the attention of all mathematicians. For example, faddism determined some directions of research. The pointlessness of some of what was done in past centuries might serve to dampen the fire of rampant current fads. History might also remind mathematicians of the major problems and goals. It is even possible that researchers, informed by scholars of the history of mathematics, might learn humility by becoming acquainted with the truly great works of the past.

Scholarship is not easy or shallow, as researchers would have us believe. It requires a mentality more precious than most original research requires, a judgment of intellectual values that enables one to discriminate between the significant and the incidental, a sophisticated and well-developed common sense, and a sympathetic imagination.

Victor F. Weisskopf, a distinguished professor of physics at the Massachusetts Institute of Technology, has expressed the need for scholars in all of the sciences. Writing in *Science* (April 14, 1972), he says:

> Another destructive element within the science community is the low esteem in which clear and understandable presentation is held. This low esteem applies to all levels. The structure and language of a scientific publication is considered unimportant. All that counts is the content; so-called "survey" articles are understandable only to experts; the writing of scientific articles or books for non-scientists is considered a secondary occupation and, apart from a few notable exceptions is left to science writers untrained in science, some of whom are excellent interpreters. Something is wrong here. If one is deeply imbued with the importance of one's ideas, one should try to transmit them to one's fellows in the best possible terms. . . . Perhaps a lucid and impressive presentation of some aspect of modern science is worth more than a piece of so-called "original" research of the type found in many Ph.D. theses, and it may require more maturity and inventiveness.

Why aren't scholars performing the functions we have described? Because scholarship, as opposed to research, is not valued, professors do not cultivate it. There is no place for the scholar in the mathematical world. He either receives no appointment or, if by chance he should acquire appointment and tenure, his work is not recognized. Nor does he find a ready outlet for his work. Current policy bars the use of space in prestigious journals unless one has something new to report.

Mathematicians have always constituted a clannish, elitist, snobbish, highly individualistic community in which status is determined, above all, by the presumed importance of original contributions to mathematics; and in which the greatest rewards are bestowed upon those who, at least in the opinion of their peers, will leave a permanent mark on its evolution. To most mathematicians scholarship is a confession of failure as a researcher, a tacit admission of inability to compete in the arena of pure mathematics. Thus, in the highly status-conscious world of mathematics it takes courage and even sacrifice to violate the canons of respectability; one incurs derision or ostracism.

A scholar performing the functions we have just described may not be a good teacher. The assimilation, synthesis, and evaluation of research are as demanding in time, energy, and special cast of mind as research itself—in fact, more so than most of the research done. True, the scholar would be more broadly informed and, in this respect, better qualified to teach. But knowledge alone does not make a teacher. What is the solution to our pedagogical problem? How are we going to get capable, dedicated men and women who will fashion the courses and write the texts that will inspire and meet the needs of students with diverse interests; spend time counseling students; prepare motivating themes and routines of discovery; know and use in their teaching attractive applications to the physical, social, and biological sciences, psychology, actuarial work, and other fields; take advantage of new teaching devices such as laboratory materials and films; stimulate thinking; study the rationale and methodology of examinations and grades; and, of course, be able to communicate with young people? Such men and women must also be concerned with what the elementary and high schools are teaching, so that college courses can be built upon what the students actually know when they enter college. The desirable men and women will have to know how much young people can understand about abstractions and what motivations will appeal to students at various ages. (See also Chapter 4.)

Clearly, we need a third class of faculty—namely teachers—to perform functions that are quite different from those of researchers and scholars. We need teachers of whom students can say, with Chaucer, "Gladly would he learn and gladly teach." Specialists may believe in the value of their subject, but their beliefs usually stop short at this point. Of course, just as scholarship feeds on research, so teachers would look to scholars for sustenance. Otherwise, teaching would become a repetition of stale, outmoded, and downright incorrect prattling. The magnitude of the problem of supplying college teachers is rarely appreciated; there are now about ten million undergraduate students in colleges and universities.

College administrators and professors accept, or pretend to accept, as axiomatic that a Ph.D. well prepared in his specialty, competent as a researcher, intelligent, and sincerely devoted to his subject will certainly be a fine teacher; preparation, training, and special qualifications for teaching are unnecessary. They hold to these beliefs despite the oft-made observation that college teaching is a major learned profession for which there should be a well-defined program to develop the skills that practitioners must possess and to select those who can acquire the skills. Many professors regard with disdain the training needed for teaching.

To interest men and women in a college teaching career, we shall have to change radically the attitudes and policies of universities. The shibboleth of research, the glorification and the superior emoluments accorded to it, must be countered if we are to combat the low esteem in which teaching is held. Teaching must be rewarded in every way as much as research. Teachers influence thousands of students during their careers and are far more vital to society than researchers except for a Newton. At present, teachers in universities, if retained at all, are tolerated, and their salaries are far below those of researchers.

The argument is often advanced that teachers stagnate. But if the functions described above are required and are performed, there would be little likelihood of stagnation. Indeed, the risk of stagnation is greater in the case of researchers, who, as we have noted (Chapter 4), can and do burn out. Since their function is to strengthen the research capacity of a department, if they do deteriorate they are truly useless for many years of their tenure.

Many people have called for recognition of scholarship and teaching. Dr. Alvin Weinberg, Director of the Oak Ridge National Laboratory, has spoken out forthrightly:

> In the first place the university must accord the generalist of broad outlook the status and prestige it now confers solely upon the specialist of narrow outlook; and in the second place, the university must rededicate itself to education, including undergraduate education. I realize that the first of these measures is viewed with suspicion by the university. Specialization is "blessed" in the sense that only the specialist knows *what* he is talking about; yet, if only the specialist knows what he is talking about, only the generalist knows why he should talk at all.*

*   *Reflections on Big Science*, M.I.T. Press, 1967.

In an article in *Science* (May 6, 1965), Dr. Weinberg stressed the need for teaching:

> The university must recognize its traditional mission—education of the young. Pregraduate education ought to give wholeness to the university. Education at the undergraduate level should properly be less professionalized and puristic than it is at the highest levels. Just as ontogeny recapitulates phylogeny, so elementary education properly should recapitulate the historic path of a discipline, its connections with other disciplines and with practical purposes—in short its place in the scheme of things. If the university takes undergraduate education seriously, and does not look upon it as attenuated professional education, the university community will be forced to broaden its outlook.

Should scholars and teachers be expected to publish? Though exceptions should be permitted, some publication is in order. However, the word *publication* must be understood in its broadest sense. It should embrace far more than research papers containing new results. A good expository paper will benefit far more people than most research papers. A good text is worth a thousand of the usual trifles that appear in research journals. Significant articles on pedagogy are sadly needed and should be welcomed.

Will all scholars and teachers perform as expected? Of course not. The variation of quality among such people will be as great as that among researchers. Some pressure to pursue self-development and up-to-dateness can be exerted by using promotion and salary as incentives. But these measures are all that one can invoke to oblige researchers to continue to produce high-quality research.

Scholars and teachers must be trained. The present staffs of graduate mathematics departments cannot do the job, and changes will have to be made. Research professors cannot teach the appropriate courses and even regard it as beneath their dignity to do so. Professors who are sympathetic to training scholars and teachers will have to be appointed, and graduate courses that stress breadth of knowledge rather than specialization will have to be instituted. Moreover, there must be collateral courses in the physical and social sciences that will enable teachers to acquire knowledge that can be used to motivate students and to teach applications of mathematics. Such professors and courses would educate college teachers and provide the proper instruction for prospective elementary and high school teachers.

The broadening of graduate education is necessary for other reasons as well. One of the new features of the current educational process is the rise of community (two-year) colleges. Some figures are highly relevant. The University of California has 100,000 students at eight *university* centers. There is a California state *college* system with 250,000 students on nineteen campuses. In addition, there are 100 *community colleges* in California that cater to 700,000 students. Nationwide, about 50 percent of the full-time undergraduate college students are in about 1,200 community colleges—as opposed to about 1,600 four-year colleges and universities—and the percentage in the community colleges is sure to increase.

The community colleges face a number of special teaching problems. A third of their courses are essentially remedial. About 75 percent of the students seek to learn only technical subjects that will prepare them for jobs, and they terminate their education at the end of two years. Such fields as computing, statistics, probability, and applied mathematics at a low level are the ones that must be emphasized. Clearly, the teachers must have a broad background, possess interdisciplinary skills, and be willing to work at the level in question.

Many community colleges are wary of hiring Ph.D.'s and even prefer to hire high school teachers. Certainly, Ph.D.'s geared to research are not the proper teachers for community colleges. Those who really want to do research will not take any interest in teaching and will leave as soon as possible for a university position, and those who are willing to devote themselves to the requisite tasks are poorly prepared. Even most Ph.D.'s who teach at four-year colleges are not properly prepared. They too are asked to teach statistics, probability, computer science, and physical applications, but they have little idea of how to cater to such interests.

Beyond training two- and four-year college teachers, graduate programs should serve other interests. The variety of graduate students, high school and college teachers wishing to improve their backgrounds, housewives returning to teaching, practicing engineers and statisticians recognizing the need for more mathematical training, and adults seeking advanced or updated education, must be catered to. History does not ensure prophecy, but the growth in technology seems to indicate an increased demand for graduate education by nonspecialists in mathematical research.

Unfortunately, the many fine intellects that could be devoted to scholarship, teaching, and the training of scholars and teachers are now hindered or even wasted through miseducation. In an article in the *American Mathematical Monthly* (September 1969), Professor I. N. Herstein of the University of Chicago pointed out that 75 percent of the students trained to do mathematics research never do so after acquiring the Ph.D.; they become teachers at two- and four-year colleges that do not demand research. Other studies confirm this fact. Moreover, surveys made of the research done by Ph.D.'s report that fewer than 20 percent published even one paper a year, and this says nothing about the quality of those papers.

No doubt many factors disincline or deter Ph.D.'s from pursuing postdoctoral research. Most of those men and women, intelligent enough to complete a doctoral program, could be trained to be scholars and teachers. But currently they are forced to take a research-oriented Ph.D. The practice of training researchers and then asking them to spend a good deal or all of their professional lives at teaching contributes to the debacle of undergraduate teaching. The universities are wasting an enormous amount of professorial and student time, energy, and resources in failing to recognize the misdirected education of the 75 percent.

Several independent commissions have urged the reorganization and broadening of graduate school objectives. One suggestion in particular has received much attention, namely a new degree, to be called Doctor of Arts, that would serve scholars and teachers. As far back as 1946, Howard Mumford Jones, in his *Education and World Tragedy*, advocated a separate degree for scholars and teachers. The Carnegie Commission on Higher Education, in *Less Time, More Options*, a report of 1971, not only supports the Doctor of Arts degree but also points out why it is far more desirable than the Ph.D. for prospective teachers. The report states:

> We consider it of great importance to reduce the impact of specialization and research on the entirety of higher education.... The Ph.D. now has a headlock on much of higher education. The greatest rewards are given to those who are highly specialized in their interests and undertake research, sometimes almost regardless of its importance and their own interest in it. The curriculum nearly all along the line is geared to the interests of the specialized instructor and to training the student for specialization. The requirement of the [Ph.D.] dissertation orients the student toward his research specialization rather than toward the instruction of students; it sometimes results in a "trained

incapacity" to relate to students by both the selection process it sets in motion and the standards of performance it imparts. We now select and train a student to do research; then employ him to teach; and then promote him on the basis of his research. This both confuses him and subverts the teaching process.

The report stresses the need for a degree program that qualifies people to become teachers, to declare by its very existence that teaching is also important and will be equally rewarded. Narrower and narrower specialization should not dominate higher education. In fact, the Doctor of Arts should be the standard liberal arts advanced degree; the Ph.D. could be retained for the relatively few who train for research. Both the Carnegie Corporation of New York and the Sloan Foundation have supported the initiation of the D.A. degree in several universities. As of 1975, about twenty-three universities, none of the most prestigious group, offer the D.A. in one or more departments.

Though the specific requirements for the D.A. degree may vary from one institution to another, it should call for a broader knowledge than the Ph.D. degree, perhaps include a course on the nature and problems of a liberal arts program, and require an expository thesis that would also evidence ability to write lucidly. The thesis might synthesize existing knowledge or even tackle many unsolved problems of education, such as motivation.

Many people believe that a Doctor of Arts degree is a simplified doctoral program, a second-rate Ph.D. It is, in fact, the Ph.D. program that is second-rate: Candidates take routine courses, show some ability to learn these in an examination that any D.A. could equally well be required to pass, and then take specialized courses or seminars that require no more than a capacity to learn. There is, of course, the thesis, which in mathematics at least calls for original research. But the problem is set by a professor, and usually the student gets continual help from the professor. Adolf Hurwitz, a leading mathematician in the first quarter of this century, rightly said, "A Ph.D. dissertation is a paper written by a professor under aggravating circumstances."

The usual thesis is not a sizable contribution to new knowledge or, if it is, is not necessarily the student's contribution. Thus, the Ph.D. does not certify research potential or promise a continuing interest in research. What is finally produced depends very much on the standards of the professors. These are often low, because some professors are too kind and others are anxious to show how many doctoral students they are turning out. The criterion of publication in a respectable journal also ensures nothing, because almost anything can be published these days—and, in fact, publication is no longer a common requirement. In many subjects, the Ph.D. thesis requirement of an original and significant contribution to knowledge is almost meaningless. Graduate students in the physical and biological sciences often work as members of a team, and the students are allowed to choose some detail of the team's results as the substance of their theses no matter who did the real thinking. As for Ph.D. theses in the social sciences and the humanities, the less said about originality and significance, the better.

On the other hand, an expository, critical, or historical thesis such as might be required for the D.A. degree could be demanded of the student as *his* work. There would be far less question of who is accomplishing what. The quality of the Doctor of Arts degree, as in the case of the Ph.D., will depend on the quality of the institutions and standards of the professors. The qualities and performance required of the candidates could be so demanding that a professor could justifiably say to an Arts degree candidate, "If you do not feel up to this program you can settle for the Ph.D."

The usual product of the Ph.D. program is a person of narrow knowledge and broad ignorance. The D.A. may not be any wiser, but he will have some breadth and will be trained for the work that he and most Ph.D.'s subsequently do.

The Doctor of Arts need not refrain from research. If he finds that he likes research, he can pursue it to the extent that time permits; he might later take a Ph.D. program, engage in research, and publish. Universities could even grant the Ph.D. to such a candidate solely on the basis of publications, which are more likely to be valuable than the hothouse-forced, professor-guided Ph.D. theses. Alternatively, the Doctor of Arts could pursue research training in an organization such as The Institute for Advanced Study in Princeton, which gives no degrees or credits but does offer courses and contact with leaders in research. Training in research could be a postdoctoral undertaking, just as specialization in a branch of medicine is post-M.D. training.

Another group, the Panel on Alternate Approaches to Graduate Education, in its report *Scholarship for Society* (Educational Testing Service, 1973), has also advocated broader programs in the graduate schools. In the light of modern needs and changing circumstances, it concluded that new elements need to be added to the graduate programs and the horizons of concern expanded. Interdisciplinary studies were also stressed. The Panel scored the graduate schools for their concentration on training only for research and publication. The faculties of the graduate schools, now under the control of research professors, were urged to take a broader view of their professional roles and not to make decisions on tenure, promotion, and salary even for *graduate* faculty only on the basis of research and publication:

> It is a matter of recreating the graduate faculty as leaders in the search for a new understanding of the possibilities of human society and of recreating the graduate institution as one that is capable of counseling political and cultural leaders on ways of assuring meaning to the structural changes of society now in progress....There are new roles as well as a new clientele for graduate education....

The Panel concluded that there must be a "re-evaluation of the basic objectives and organization of each graduate school and the disciplines essential to it."

President Derek C. Bok of Harvard University, in his annual report of 1972-73 to the Board of Overseers, stated quite frankly that Harvard's Graduate School of Arts and Science

> has not done enough to prepare its students for the role they will play as teachers and educators.... Yet the fact remains that most of our graduate students obtain their doctorates without ever being exposed to the best of the literature on higher education and without ever exploring systematically the educational issues they will encounter as teachers, committee members and department chairman.... This situation is hard to defend. Students seek graduate training to pursue specific careers, yet graduate schools make a serious effort to prepare students for only a part of their professional lives. This practice is hardly suited to an age when society increasingly questions the quality of college teaching and the value of traditional forms of higher education.

President Bok agrees with other surveys that more and more of even Harvard Ph.D.'s will be devoting their careers to teaching and adds, "But surely it is odd to continue placing such exclusive emphasis on research when so many of our students will spend large parts of their careers in predominantly teaching institutions." He also points out that the problem is hardly new and has been raised many times in the past by presidents, deans, and iconoclastic professors.

Yet little has changed, either at Harvard or elsewhere. There is a good chance, however, that the universities will broaden their graduate school programs in the near future. But if they do, in most cases it will not be because they wish to make amends for the narrowness of their present programs or because they have become conscience-stricken about their indifference to the needs of society. It will be to alleviate financial problems by admitting more students and collecting more tuition.

The current academic effort is dichotomized. When research in the United States became a viable activity, the universities, falling for the prestige of research and financial gains made possible thereby, strained their manpower and other resources to win that prestige. But the country as a whole, dedicated to universal education at higher and higher levels, needs broadly informed and competent teachers and scholars—which it has not yet produced. Just as two men pulling on opposite ends of a rope fall to the ground when the rope splits, so our mutually competitive educational efforts have suffered. The research has sunk to fruitless specialization and our classrooms on all levels are staffed with poorly trained or mistrained teachers. The empty space that separates the two ends of the split rope is, in the educational area, the vacuum that remains to be filled by teachers and scholars.

The problem of what to do with research professors who cannot teach even on the graduate level also remains to be solved. Actually, in some cases teaching and research could complement each other; some professors do derive stimulation from teaching. Moreover, teaching can keep a professor psychologically at ease during an otherwise unproductive period, because the time is being profitably used. Further, it is the desire of any professor who has a worthwhile direction of research to see it pursued, and his students are the most likely candidates for such pursuit. But researchers who cannot or will not teach should be employed in institutes that are devoted entirely to research, conduct no formal teaching, and give no degrees. In such institutions the researchers would be free to pursue their interests full time.

The benefits to be derived from segregating excellent researchers who cannot or will not teach are immense. To burden such people with formal teaching is to deprive them of time and energy and to distract them from their valuable work. It is equally important that such researchers be dislodged from dominance in formulating university policies. Although they may soar to great heights of thought much as a towering mountain peak may rise among hills, they are limited in range. Their role in the educational process may be detrimental. As long as researchers are regarded as the cream of the university personnel, more valuable than scholars, teachers, and administrators, their needs and judgments will usurp those of others. The elitism and snobbery of some are an affront to sincere and capable colleagues. Recognition of the shortcomings of researchers is as important as helping them to function effectively.

Special institutes for researchers abound in Germany and the Soviet Union, where they are government-supported. In Germany, for example, the Max Planck Gesellschaft (the Max Planck Society for the Advancement of Science) directs about fifty research institutes called Max Planck Institutes. They employ about ten thousand people and cover many areas of research. (The Max Planck Gesellschaft is the direct successor of the Kaiser Wilhelm Gesellschaft, which was founded in 1911. One of the institutes, devoted to physics, was founded for Albert Einstein in 1917.) These institutes are free to choose their own direction of research. Though there are government-supported laboratories in the United States, these have specific missions and are not free to determine their research programs. The Institute for Advanced Study in Princeton* is the only organization that approximates the German and Soviet institutes; however, our own Institute is privately endowed and cannot be the haven for all who should be accommodated. This is not to belittle its contribution. For instance, when Einstein was forced out of Germany by Hitler he became one of the six chosen to inaugurate the mathematics division. Certainly, a number of such institutions for fine research people would keep research active and permit universities to confine themselves to researchers who would promote the numerous functions of the graduate schools.

Still another measure could be adopted by some universities. A few declare openly that their chief function is to do research and train research people. Since the faculties of these universities either cannot teach undergraduates or, even if capable of doing so, cannot take the time and energy to do so properly because research is so demanding, these universities should discontinue undergraduate education. Of course, they would be reluctant to do so. They want the undergraduate courses to help provide jobs for the researchers and for graduate students, and they want the tuition that undergraduates pay, or that the states provide, to support research. But it is just these abuses of undergraduates that must be eliminated, and any justification offered by these universities for retaining undergraduate schools (such as the opportunity they give young students to come into

* The Institute has no formal connection with Princeton University. It has its own administration, objectives, and funds.

contact with "great" minds) is factitious. The model for purely research-oriented universities could very well be Rockefeller University in New York City. It is outstanding in research strength and offers only Ph.D. degrees.

However, there are measures that can enable a university to educate undergraduates and still retain the research and graduate training. To eliminate robbing undergraduates in order to favor graduate teaching and research, the two missions of mathematics departments (and other departments)—graduate and undergraduate—should be administered separately and their budgets maintained separately. Undergraduate tuition should be used solely for undergraduates. Moreover, appointments to undergraduate teaching should *not* be subject to approval by the head of the graduate department. The segregation of funds and authority would permit the undergraduate divisions to hire full-time competent teachers in place of the low-paid, immature graduate students who now do the bulk of undergraduate teaching. Any surplus could be used to increase scholarships and facilities.

The suggestion that the undergraduate and graduate functions be separated may appear radical. Actually, it would be a reversion to an older order. When graduate education and research were first undertaken in this country, during the last quarter of the nineteenth century, the graduate faculties were separate. It was President Charles W. Eliot of Harvard who, in 1890, first established a single faculty for both divisions. This move was wise *at that time*, because the teaching faculties were weak and the researchers were more able teachers of the advanced courses. Also, through close contact with the teachers, the researchers exerted a largely beneficial influence.

The separation recommended for today's universities does not mean that one division should not influence the other, or that the faculties must be entirely distinct from one another. A wise undergraduate chairman will utilize the services of a graduate professor if he believes it will improve the undergraduate program, and the graduate chairman will do the corresponding thing. A wise dean would suggest or encourage such cooperation.

Researchers oppose the total separation of graduate and undergraduate schools. The arguments are, first, that "A good graduate school provides the best way of assuring high academic and intellectual standards." Just how these standards work their influence on the undergraduate schools is not made clear. The research professors are not at all interested in the undergraduate activity. What is clear is the disregard of the undergraduates in favor of the graduate faculty (Chapter 5). Second, "A good graduate school provides the best way of recruiting a distinguished faculty. The strength of a faculty depends on its creativity—on the opportunity to do research, write books and work with advanced students." Again, what does this have to do with undergraduate teaching?

And then comes the giveaway. "Third, the graduate school is responsible in great part for the reputation of a university, *for the impression outsiders have of it*" (italics added).* But what value does the reputation of a university for graduate work or the impression this reputation creates have for undergraduate teaching? And if one does attract good undergraduates but does not do justice to them, how can one justify attracting the students and then fouling up their lives? The prestige of the graduate school ensnares the best undergraduates, springs the trap, and then lets them perish while struggling to free themselves. One is quite safe in assuming that the more prestigious the university, the

---

*   These arguments were presented, as quoted above, on the Op-ed page of the *New York Times* of December 17, 1975.

fewer its educational concerns and the poorer its educational effectiveness. Certainly, as long as research is the criterion of faculty value, few benefits will accrue to teaching.

Still another argument for the status quo is that graduate schools "attract the best teachers." If the purport is that the presence of a graduate school attracts good undergraduate teachers, then the facts deny this contention. A good graduate school attracts young people who seek to promote their own research more readily, and these people aspire to become prestigious researchers rather than great teachers. Moreover, at most universities they have no choice; their futures depend only on their achievements in research.

The claim that undergraduate professors need the backing, stimulus, and knowledge of the researcher is false. There are some excellent four-year colleges in this country where the best teaching takes place (Chapter 5). The professors in such schools are entirely isolated from research professors. A testimonial to the superior teaching performance of the four-year colleges was recently given by a professor at one of the most prestigious universities. Speaking of his own institution he said, "It wants to be Amherst to its undergraduates and Yale to its graduate students." Surely, then, independent undergraduate divisions in universities would not suffer.

The objection that research professors advance to the separation of graduate and undergraduate colleges of a university is entirely self-serving. Political colonialism of the nineteenth century has been refashioned into academic colonialism by which the graduate division preys on the undergraduate just as surely as wolves prey on sheep.

If the graduate schools are forbidden to use undergraduate funds, how will they maintain themselves? The answers are to cut down the number of research professors, require more teaching, cut out the competition for high-salaried researchers, and eliminate those men and women who, though called professors, contribute no more than their names. Money would also be saved on supporting personnel and space. The indirect gains would also be appreciable. Fewer specious or trivial papers would be published, and the truly worthy ones would stand forth. Further, because the number of journals in all fields is now so large that libraries cannot find the space and funds for them, reduction in the size and number of journals would reduce library costs. These costs are now so great that a current study financed by major foundations (*New York Times*, June 23, 1975) is seeking new ways of preserving and disseminating research that would eliminate journals.

Research and graduate teaching were dire needs in 1876. In 1976 both must be re-examined. While a greater variety of graduate programs is needed, the total number of graduate degrees can be reduced now and for the forseeable future; all predictions of undergraduate enrollment agree that it will decline. The universities are even now competing fiercely for students, and the competition will be stiffer as college enrollment drops. Quality of education is in the long run the best attraction, whereas the roseate bubble of prestige, which has enveloped many institutions and enabled them to look attractive, may be pierced and collapse. In any case, the sacrifice of the undergraduates to favor the graduate activities cannot be tolerated.

The separation of undergraduate and graduate functions would have another positive value. At present the running of the undergraduate activities—decisions on courses to be offered, selection of teachers (whether graduate students or mature people), choice of texts, and teaching schedules—is assigned to an administrative assistant by the head of the department. The low quality of the resulting undergraduate education does not worry the head. He knows that his performance will be judged by the research activity. On the other hand, the undergraduate chairman, if properly chosen and a power in his own right, would seek out competent teachers. Thus, teachers would finally find a place in the universities, and the money to pay them their due would also be available. The graduate departments, in turn recognizing that their existence would depend in part on getting jobs for their graduates, might be obliged to introduce a doctoral program that would actually train teachers.

Of course, there are disadvantages and even dangers in separating graduate and undergraduate administrations and budgets. Friction, competition for funds and facilities, and even opposition may develop. It is trite to say that we must choose the lesser of two evils, but when the lesser evil is to do justice to the millions of undergraduates and, at that, indirectly provide some benefits to graduate teaching and research, the choice is clear.

It is not unlikely that even government intervention may force improvements. The most prestigious universities practiced racial and religious discrimination for decades. Present federal regulations have compelled elimination of this and other unfair practices. The likelihood of further governmental action is implied by the statement made by President Derek C. Bok of Harvard in his 1972-73 annual report to the Board of Overseers: "If universities accept huge sums in federal aid for research and training, public officials could not fail to pay attention to the way in which tax dollars were spent." State departments of education have been rather lax in imposing standards of quality, but the call for a crackdown has already appeared in newspaper editorials.

Americans might well consider another measure. Elementary school teachers used to receive training in what were called normal schools, and secondary school teachers were taught in colleges devoted to teacher training. Both types of schools were indeed poor, but only because all levels of education and knowledge in this country were poor. By 1960, the normal schools and teachers colleges had converted to liberal arts colleges, often with graduate and research programs. The latter have sought to emulate the large universities and have enlisted faculty who are subject-oriented and who seek to obtain recognition in their respective professional circles. Teacher education has been sacrificed. Prospective teachers now attending these altered colleges or the liberal arts colleges of the universities are taught content by the typical Ph.D. or by a graduate student, neither of whom is informed in the specialized needs of the students. We might now reverse history and re-establish four-year colleges devoted to teacher training. Such colleges today could demand and obtain knowledgeable faculty that would devote itself to the education of teachers. Perhaps the wisdom of such a move might be more evident by considering the situation in engineering schools. These schools must be separated from liberal arts colleges because the engineering students have to learn many technical subjects. Nevertheless, in most universities all mathematics courses are taught by liberal arts professors who care little and know less about what kind of mathematics is best for engineers. Hence, as noted earlier (Chapter 7), many engineering colleges of the large universities offer their own mathematics courses rather than have their students take those offered by the academic mathematics department. Though the proper education of prospective school teachers is closer to liberal arts than is that of future engineers, nevertheless there are features that demand specially oriented professors.

The need for such specialists is underscored by events of the 1960s. College professors should know what students learn in high school, and professors who train prospective high school teachers should certainly know what these teachers will face. During the 1960s the high schools most commonly taught the New Math. When the subject of the New Math was broached in conversations among professors, even men who were specifically undergraduate teachers would ask, What is the new math? Whether it was good or bad, the high school teachers were obliged to teach it. Certainly, then, college professors involved in training high school teachers should have been fully informed about that curriculum. Indeed, many apologists for the New Math blame its failure not on the content but on the fact that the teachers were not prepared to teach it. There is some substance to this contention.

Since the universities do not on their own initiative undertake to meet the needs of most of the students, one might look to other agencies to exert pressure on the universities. The most prestigious agency is undoubtedly the National Academy of Sciences. The Academy should certainly be concerned with the scientific health of the nation. But apparently, the most important function of this organization is to elect members—or, as one member put it, to bore those who belong and to embarrass those who do not. The election process itself may give some indication of what can be expected of the Academy. Election must be initiated through nomination by a member. Whom would a member know? Most likely his own colleagues. And he would prefer such men because honored colleagues imply honor to oneself. On the other hand, a rival in research is clearly not worthy of honor. The actual election of nominees is a political bargaining session. Because the members, coming from numerous, diverse, and highly specialized fields, really cannot judge most of the men they vote for or against, many biases determine the decisions.

The election process is incredibly complex and devious.

When the academicians Norbert Wiener and Richard
Feynman, both distinguished in their respective fields of
mathematics and physics, witnessed these elections, they
resigned in disgust. The topologist Stephen Smale, after
attending his first Academy meeting, remarked irreverently,
"It was really fantastic as these days passed to see how this
group of America's most celebrated scientists meeting
together could be so dominated by the question of just how
to increase their membership and ways to remember their
dead." The outcome of elections is that the in-group
perpetuates itself. This is also evident from the fact that no
engineers, medical people, or social scientists were among
the Academicians until about ten years ago when, after
protests, the Academy decided to elect members from these
fields. It is significant that in the years between 1950 and
1973 twelve American scientists who won Nobel prizes—
surely, some deserved it—were not members of the
Academy at the time. On the whole, the membership
consists of researchers who do not possess breadth of vision;
one would not, therefore, expect the Academy to help in
pedagogical matters. But one can expect the members to be
concerned about the enormous waste in research and the
damage that the flood of publication is causing. However,
the Academy does nothing about these problems.

Once in a while the members are stirred into action and
appoint a committee to perform a particular function. In the
late 1960s, when the United States government was still
pouring money into mathematical research, the members of
the Academy were not satisfied that the amount was enough
and appointed the Committee on the Support of Research in
the Mathematical Sciences, known as COSRIMS. The
Committee decided that its best argument for increased
support would be that the nation needs more Ph.D.'s for
college and university teaching. Even by 1965, independent

studies of future college populations were estimating that the college population would stabilize by 1970 and, somewhat later, decrease. These estimates were reliable, because the number of children who would be of college age in the next two decades was known, and the proportion of high school students entering college had not changed appreciably in several decades. Moreover, the mathematics departments were already turning out Ph.D.'s in greatly increased numbers. In fact, Dr. Allan Cartter, at that time head of the American Council on Education, warned the academic community during the boom years of the mid-1960s that there would be an oversupply of Ph.D.'s and this was evident, some said, even in 1961.

Nevertheless, the figures issued by COSRIMS indicated that there would be an enormous shortage of Ph.D.'s in the 1970s. In fact, the Committee recommended an increase of two hundred more in mathematics each year over the preceding year for the next five years after 1968. Because students were thus led to believe that jobs would be available, the number of Ph.D.'s awarded in mathematics increased from 970 in 1968 to 1,281 in 1972. It has since begun to drop, because the shortage of jobs has discouraged students from entering the field. As one article put it, the good news in the summer of 1974 was that unemployment among Ph.D.'s was not much worse than in 1973.

COSRIMS not only did not do its homework but allowed its desire to obtain more funds for research to be defended on simple-minded extrapolation from what was happening in the 1960s. In effect, as one prominent professor observed, the report was a piece of propaganda rather than one based on hard-headed research. The work of COSRIMS was but one activity of the National Academy, and one might be tempted to dismiss this one "mistake" as minor. But those young people who were encouraged to take a Ph.D. in the expectation of obtaining employment as college teachers and are now walking the streets cannot be quite so charitable.*

Another organization that could have significant effect on unbridled, low-quality research and on the programs of the graduate schools is the American Mathematical Society. At its founding in 1894, the Society planned to devote its efforts to teaching, writing, research, and any labor that any member might consider appropriate (Chapter 2). Writing, on all levels, was explicitly recommended. In particular, the Society described itself as an organization of teachers: "Any tendency to restrict its usefulness solely to the paths of investigation [research] and publication should, for every reason of prudence and wisdom, be resisted." Of course at that time research in the United States was in its infancy. As research became more active, the Society turned its attention more and more in that direction, and it began to extol research as the only significant activity of college professors.

One would expect, however, that the interest in good research and in the diffusion of relevant knowledge into the classroom would cause the Society to concern itself with teaching and expository writing and to combat overspecialization, the flood of poor research articles, the imbalance between pure and applied mathematics, and the ever increasing isolation of mathematics. But none of these evils has been tackled.

The measures that the American Mathematical Society could adopt to check excessive and wasteful publication, to aid teaching, and to reduce the inhumane pressure on young Ph.D.'s to publish would require careful consideration. A

---

*   The shortcomings of the Academy are described in Boffee, Phillip M.: *The Brain Bank of America*, McGraw-Hill, 1975.

radical measure, but one that seems to have at least more merits than defects, would be to prohibit men and women who are only two or three years past the Ph.D. from publishing in its journals. This not only would permit the young professors to adjust themselves to teaching but also would eliminate many of the worthless articles that are now flooding the journals. And it would oblige administrators to judge people rather than to count publications.

Certainly, the Society should advocate a wiser appointment policy. In place of the widespread three-year appointments, new Ph.D.'s should be appointed with no time limitation other than the understanding, as recommended by the American Association of University Professors, that they must serve seven years to acquire tenure, with notification at the end of the sixth year if tenure is not to be granted. Naturally, those who are *clearly* unsuitable could be dismissed sooner because the appointments are legally on a year-to-year basis. But if a young teacher feels that he has at least six years in which to adapt himself to a faculty position and can therefore allow some time to recognize his natural interests and aptitudes, he is more likely to devote himself to teaching, to building a better background for research, and to determining what his major contribution to the university can be. A new Ph.D. also needs time to recoup psychologically from his doctoral work. The university in turn will have more time to judge what his actual strengths are. If it should be necessary to terminate the appointment of a teacher who has served for five or six years, such a teacher would normally be about thirty years old, still young enough to find another position. In fact, having had the opportunity to examine himself, he could seek an appointment suited to the career most congenial to him.

The Society could expose other undesirable practices. It could, for example, ask for more integrity in university administrators. Many a department today appoints dozens of new Ph.D.'s for a three-year period knowing that there would be no possibility of a permanent appointment for more than one or two. These appointments are not carefully screened. The reason given for the multiple appointments is that the tyros will be observed during the three-year period and the best retained. But the real reason for the policy is that the university can boast of having a large number of research-oriented Ph.D.'s while paying them about one-third of a full professor's salary.

The Society also should be concerned with the quality of its publications. Since refereeing of papers is now an almost impossible task (Chapter 3), the Society could insist that papers submitted to its journals be readable; thus, at least correctness could be determined. Most authors write as though they were more concerned to be admired than understood. The full worth of a paper may be harder to determine, but even this judgment could be facilitated. Each author should be required to state what, at least in his opinion, is the contribution of his paper—application to science, aesthetic value, or intellectual interest. The most likely objection to such requirements is that more journal space would be needed, and journal space is expensive. But the proposed requirements would result in eliminating a great many of the presently published papers because they are either incorrect, worthless, or duplications. The net result would be a saving of space and easier determination by readers of whether a paper is relevant to their interests. With the present uncontrolled flood, the good is swamped by the bad.

Still another organization, the Mathematical Association

of America, explicitly states its devotion to furthering the interests of undergraduate education. This organization should certainly be fighting the overemphasis on research, the low status accorded to teaching, the lack of training of graduate students for future teaching, and the use of teaching assistants. In fact, the organization was founded by a group of the American Mathematical Society, which in 1915 recognized that the Society had begun to ignore teaching in favor of research. But this organization is no more effective in its sphere than the Society is in research. No more can be expected of it than to maintain the status quo.

Both organizations appoint committees that report on current problems and urge the appointment of more committees and the gathering of information. In the last few years the need to diversify the training of mathematicians has been stressed by panels at meetings and occasional articles in the journals of the two societies. A recently appointed committee even pokes a little fun at one organization by using the title "How to Cope with M.A.A.: Mathematical Avoidance and Anxiety." But a scrutiny of these talks and articles soon yields the underlying reason for urging any reform—to help Ph.D.'s get jobs. For example, a rapidly expanding possible market for Ph.D.'s is the two-year junior college. Hence, the panels and articles address themselves to the needs of these colleges. Industry also needs mathematicians, and applied mathematics is frequently recommended as preparation for such jobs.

Of course, the advice from present-day professors in universities on the needs and problems of the junior colleges and industry is almost worthless. Very few of the men have had experience in either field and so do not know what these organizations require of employees. The advice expresses pious sentiments motivated solely by the job shortage. The overemphasis on research, the narrowness of the Ph.D. training, and the miserable teaching of undergraduates in most universities are not attacked.

These articles and speeches attempt to cure a serious illness with aspirin. No concrete, effective measures are undertaken. Actually, many members of both societies are influential in their own university departments; some are chairmen. But they do nothing at their own institutions to broaden the nature or goals of graduate education or to secure and retain scholars and teachers. Research is the coin of the realm. Ironically, some of these chairmen sought administrative positions because they did not do research. Though a few chairmen now favor a broader thesis that might be an expository dissertation rather than original research, in the present atmosphere the chances for a person who writes such a thesis to get a job are practically nil. One chairman magnanimously said he would hire such a person for two years, but of course would not keep him beyond that period.

One other organization, the National Council of Teachers of Mathematics, warrants mention. This organization is concerned with high school and elementary school education. It is more effective in its sphere than either of the other two. High school and elementary school teachers do not have the time and opportunity to develop and test new ideas, new techniques of teaching, and new communication media. Through its journals, meetings, and workshops the National Council does provide instruction in content and methodology. But some of its leaders have been gullible. In 1961, before the New Math had received any significant testing, the Council published a pamphlet, "The Revolution in School Mathematics," that not only advocated the New

Math but charged that those schools that failed to adopt it were remiss in their obligations.

Moreover, the Council should represent the views of all the members. It has not always done so. For example, during the heyday of the New Math, the Council's journal, *The Mathematics Teacher,* could not avoid publishing some articles condemning the New Math. Thus, when a critical speaker, invited to give a keynote speech at some meeting, submitted his speech for publication, the editors dared not reject it. But they made sure it was followed in the same issue by a rejoinder. Apparently the editors wished to prevent criticism of the New Math from having any influence on its members. Most editors of the *American Mathematical Monthly,* the official organ of the Mathematical Association of America, also reject articles critical of the existing conduct of education.

The cause of the shortcomings in the several organizations is readily explained. Leadership in them offers prominence, prestige, and—indirectly—personal advancement in one's university, college, or other institution. Hence, such positions are sought by members of the profession. How does one attain high position in these organizations? The procedure is not unlike what happens in politics. The president appoints the important committees, including the committees that nominate future officers and editors of journals. The committees are sure to be composed of insiders, and they make the choices for lesser committees. An aspirant to office seeks and makes the acquaintanceship of some prominent members of the society and at an opportune time offers to serve on some minor committee. These prominent members, usually already on important committees, will choose this man because his name stands out as opposed to the thousands of names of unknown members. Going along with the majority and continuing cultivation of prominent members leads to more and more important roles, visibility in society affairs, and almost inevitably to higher and higher office. Finally, the organization sends out an election ballot to its members with Mr. Blank as the only choice for president. Of course, Mr. Blank is elected. The leadership perpetuates its own kind.

That mediocrities and power- or prestige-seekers should rise to the presidency of a national organization may seem incredible. But one has only to think of some of the men who have become presidents of the United States.

Occasionally, a fine researcher is honored for his contributions by election to the presidency of one of these organizations. But it is rare that such a person is deeply concerned about the problems of the mathematical community or, if he is, that he has the administrative capacity to be effective. In fact, most presidents are content to bask in the sunshine of the office and perform perfunctorily during the year or two that they serve. In their behalf one should add that they are not relieved of their normal duties at their universities and so cannot devote much time and energy to organizational work.

However, thoughtful, sincere, and unusually energetic men can be found in the organizations we have just discussed; and if perchance they should come to the fore, they could get their respective agencies to grapple with the problems of research and teaching and certainly aid immeasurably in solving them. This does happen—but all too rarely.

The above recommendations will not solve all the problems of mathematics education. We must know more about how children learn and what motivates young people at various ages, and we must devise and employ more pedagogical aids; however, these measures call for more study and, unlike the ones recommended, cannot be employed at once.

Reforms are needed not only to improve mathematics education. Both the survival of mathematics in the curriculum and of research itself are at stake. The concentration on pure, esoteric studies will ultimately mean less support from society and, as Richard Courant once predicted, all significant mathematics will be created by physicists, engineers, social scientists, and schools of business administration. Poor education drives away students, and fewer students means fewer jobs for mathematicians. The demand for mathematics education was in sharp decline during the 1930s, and deservedly so. Only World War II revived it. One would hardly hope for a repetition of such a saving measure.

Though reforms of various sorts have yet to be adopted, we need not be too despondent about our educational system. The defects are somewhat excusable in the light of the early handicaps. As the United States grew in technology, power, and world influence, we sought to match other powers in scientific strength. We were the *nouveaux riches* who wanted scientific status. And so we imitated the western European countries, which had produced the best research. In our zeal to equal them we lost sight of one of our principles that should truly have given us pride and gratification: Unlike the European countries, we pledged ourselves early in our history to universal education and to free or low-cost education at higher and higher levels. The tasks of building and improving the educational system and at the same time striving for equality or even leadership in research were beyond the country's resources. We have attained that leadership, at the expense of quality education for all students.

It is now time for mathematicians to shed their narcissism, broaden their vision and interests, limit research to worthwhile papers, and thereby release time and energy to the many-sided needs and tasks of education. They must cater to the interests of all students, undergraduate and graduate alike. Only recognition of the interdependence of research, scholarship, and teaching can advance mathematics itself, improve teaching, and further the multitudinous valuable uses of mathematics in our society.

272   *Morris Kline*

Bibliography

Alder, Henry L. "Mathematics for Liberal Arts Students." *The American Mathematical Monthly* 72 (1965), 60-66.

Allendoerfer, C. B. "The Narrow Mathematician." *The American Mathematical Monthly* 69 (1962), 461-69.

Ashby, Eric. *Adapting Universities to a Technological Society*. San Francisco: Jossey-Bass, Inc., 1974.

Axelrod, Joseph. *The University Teacher as Artist*. San Francisco: Jossey-Bass, Inc., 1973.

Baker, Liva. *I'm Radcliffe! Fly Me! The Seven Sisters and the Failure of Women's Education*. New York: Macmillan, 1976.

Barzun, Jacques. "The Cults of 'Research' and 'Creativity'." *Harper's Magazine*, October 1960, 69-74.

———. *Teacher in America*. New York: Doubleday and Co., 1954.

———. *The House of Intellect*. New York: Harper and Bros., 1959.

———. *The American University*. New York: Harper and Row, 1968.

Bell, Eric T. Review of *The Poetry of Mathematics and Other Essays*. *The American Mathematical Monthly* 42 (1935), 558-62.

Berelson, Bernard. *Graduate Education in the United States*. New York: McGraw-Hill, 1960.

Bidwell, James K., and Clason, Robert G. *Readings in the History of Mathematics Education*. Reston, Va.: National Council of Teachers of Mathematics, 1970.

Blanshard, Brand, ed. *Education in the Age of Science*. New York: Basic Books, 1959.

Boffey, Philip. *The Brain Bank of America*. New York: McGraw-Hill, 1975.

Bylinsky, Gene. "Big Science Struggles with the Problems of its own Success." *Fortune*, July 1977, 61-69.

Caplow, Theodore, and McGee, Reece J. *The Academic Marketplace*. New York: Basic Books, 1958.

Carnegie Commission on Higher Education. *Less Time, More Options*. New York: McGraw-Hill, 1971.

Carnegie Foundation for the Advancement of Teaching. *More Than Survival*. San Francisco: Jossey-Bass, 1975.

Carrier, George F., et al. "Applied Mathematics: What Is Needed in Research and Education." *SIAM Review* 4 (1962), 297-320.

Cole, Jonathan R., and Cole, Stephen. *Social Stratification in Science*. Chicago: University of Chicago Press, 1973.

Conference Board of the Mathematical Sciences. *Overview and Analysis of School Mathematics, Grades K - 12*. Washington, D.C.: 1975.

_____. The Nacome Report. *Undergraduate Mathematical Sciences in Universities, Four-Year Colleges and Two-Year Colleges,* 1975-76. Washington, D.C.: 1976.

Duren, W. L. Jr. "Are There Too Many Ph.D.'s in Mathematics?" *The American Mathematical Monthly* 77 (1970), 641-46.

Eble, Kenneth E. *Professors as Teachers*. San Francisco: Jossey-Bass, 1974.

Educational Testing Service. *Scholarship For Society*. Princeton, N.J.: Educational Testing Service, 1973.

_____. *A Survey of Mathematical Education*. Princeton, N.J.: Educational Testing Service, 1956.

Eliot, Charles W. *A Turning Point in Higher Education*. Cambridge, Mass.: Harvard University Press, 1969.

Federal Security Agency, Office of Education. *Toward Better College Teaching*, Bulletin #13. Washington, D.C.: U.S. Government Printing Office, 1950.

Freudenthal, Hans. *Mathematics as an Educational Task*. Boston: Reidel Publishing Co., 1973.

Gaff, Jerry G. *Toward Faculty Renewal*. San Francisco: Jossey-Bass, 1975.

Gaston, Jerry. *Originality and Competition in Science*. Chicago: University of Chicago Press, 1973.

Glenny, Lyman A., et al. *Presidents Confront Reality, From Edifice Complex to University Without Walls*. San Francisco: Jossey-Bass, 1975.

Griffiths, H.B., and Howson, A.G. *Mathematics: Society and Curricula*. Cambridge, England: Cambridge University Press, 1974.

Hayes, John R. "Research, Teaching and Faculty Fate." *Science* 172 (1971), 227-30.

Henderson, Algo D., and Jean Glidden. *Higher Education in America*. San Francisco: Jossey-Bass, 1974.

Henrici, Peter. "Reflections of a Teacher of Applied Mathematics." *Quarterly of Applied Mathematics* 30 (1972/73), 31-39.

Henry, David D. *Challenges Past; Challenges Present*. San Francisco: Jossey-Bass, 1975.

Herstein, I.N. "On the Ph.D. in Mathematics." *The American Mathematical Monthly* 76 (1969), 818-24.

Highet, Gilbert. *The Art of Teaching*. New York: Vintage Books, 1955.

Hohenberg, John. *The Pulitzer Prizes*. New York: Columbia University Press, 1976.

Howson, A.G., ed. *Developments in Mathematical Education*. New York: Cambridge University Press, 1973.

Hutchins, Robert M. *The Higher Learning in America*. New Haven: Yale University Press, 1936.

Jencks, Christopher, and Riesman, David. *The Academic Revolution*. New York: Doubleday and Co., 1968.

Kendall, Elaine. *An Informal History of the Seven Sister Colleges*. New York: G.R. Putnam's Sons, 1976.

Klamkin, Murray S. "On the Ideal Role of an Industrial Mathematician and Its Educational Implications." *The American Mathematical Monthly* 78 (1971), 53-76.

Kline, Morris. "Freshman Mathematics as an Integral Part of Western Culture." *The American Mathematical Monthly* 61 (1954), 295-306.

_____. "Mathematical Texts and Teachers, A Tirade." *The Mathematics Teacher* 49 (1956), 162-72.

_____. *Mathematics in Western Culture*. New York: Oxford University Press, 1953.

_____. *Mathematics and the Physical World*. New York: T.Y. Crowell Co., 1959.

_____. "A Proposal For the High School Mathematics Curriculum." *The Mathematics Teacher* 59 (1966), 322-30.

_____. "Intellectuals and the Schools: A Case History." *Harvard Educational Review* 36 (1966), 505-11.

_____. *Mathematics For Liberal Arts*. Reading, Mass.: Addison-Wesley Publishing Co., 1967.

_____. "Logic Versus Pedagogy." *The American Mathematical Monthly* 77 (1970), 264-82.

_____. *Why Johnny Can't Add: The Failure of the New Math*. New York: St. Martin's Press, 1973.

Langer, Rudolf E. "The Things I Should Have Done, I Did Not Do." *The American Mathematical Monthly* 59 (1952), 443-48.

Levine, Arthur, and Weingart, John. *Reform of Undergraduate Education*. San Francisco: Jossey-Bass, 1973.

Lewis, Lionel S. *Scaling the Ivory Tower*. Baltimore: The Johns Hopkins University Press, 1975.

Livesey, Herbert. *The Professors*. New York: Charterhouse, 1975.

Lyons, Gene. "The Higher Illiteracy." *Harper's Magazine*, September 1976, 33-40.

Martin, Everett Dean. *The Meaning of a Liberal Education*. Garden City Publishing Co., 1926.

Mathematical Education in the Americas. Teachers College, Columbia University, New York: 1962.

Merton, Robert K. *The Sociology of Science*. Chicago: University of Chicago Press, 1973.

Miller, Richard I. *Developing Programs For Faculty Evaluation*, San Francisco: Jossey-Bass, 1974.

_____. *Evaluating Faculty Performance*. San Francisco: Jossey-Bass, 1972.

Milton, Ohmer. *Alternatives to the Traditional: How Professors Teach and How Students Learn*. San Francisco: Jossey-Bass, 1972.

National Science Board. *Graduate Education, Parameters For Public Policy*. Washington, D.C.: U.S. Government Printing Office, 1969.

Nevanlinna, Rolf. "Reform in Teaching Mathematics." *The American Mathematical Monthly* 73 (1966), 451-64.

Newsom, Carroll V. "The Image of the Mathematician." *The American Mathematical Monthly* 79 (1972), 878-82.

Nisbet, Robert B. *The Degradation of the Academic Dogma, The University in America, 1945-1970*. New York: Basic Books, 1971.

Northrop, E.P. "Mathematics in a Liberal Education." *The American Mathematical Monthly* 52 (1945), 132-37.

Ohmann, Richard. *English in America*. New York: Oxford

University Press, 1976.

Paige, L.J. "Public Understanding of Science and Its Implications For Mathematics." *The American Mathematical Monthly* 78 (1971), 130-42.

Parker, Gail Thain. "While Alma Mater Burns." *The Atlantic Monthly*, September 1976, 39-47.

Peterson, R.B. "Survival for Mathematicians or Mathematics." *The American Mathematical Monthly* 79 (1972), 70-76.

Rogers, Hartley, Jr. "The Future of the University in Mathematics Education." *The American Mathematical Monthly* 82 (1975), 211-18.

Rudolf, F. *The American College and University*. New York: Knopf, 1968.

Stone, Marshall: "Mathematics and the Future of Science." *Bulletin of the American Mathematical Society*. 63 (1957), 61-76.

_____. "The Revolution in Mathematics." *The American Mathematical Monthly* 68 (1961), 715-34.

Ulich, Robert, ed. *Three Thousand Years of Educational Wisdom*. Cambridge, Mass.: Harvard University Press, 1954.

von Neumann, John: "The Mathematician" in R. B. Heywood: *The Works of the Mind*, University of Chicago Press, 1947, 180-196; also in J.R. Newman: *The World of Mathematics*,Vol. 4, 2053-2063, Simon & Schuster, 1956.

Weinberg, Alvin M. "But Is the Teacher Also a Citizen?" *Science* 149 (1965), 601-6.

_____. *Reflections on Big Science*. Cambridge, Mass.: M.I.T. Press, 1967.

Weyl, Hermann: "A Half-Century of Mathematics." *The American Mathematical Monthly*, 58 (1951), 523-553.

Whitehead, Alfred North. *Science and the Modern World*. Cambridge, England: Cambridge University Press, 1926.

_____. *Essays in Science and Philosophy*. New York: Philosophical Library, 1948.

_____. *The Aims of Education and Other Essays*. New York: Macmillan, 1929.

Willcox, A.B. "England was Lost on the Playing Fields of Eton." *The American Mathematical Monthly* 80 (1973), 25-40.

Wilson, Logan. *The Academic Man*. New York: Oxford University Press, 1942.

Wilson, Robert C., et. al. *College Professors and Their Impact on Students*. New York: John Wiley and Sons, 1974.

Index